TECHNICAL DRAWING
BUILDING DRAWING

S.C.O.A. Ezeji

G.I. Nwoke

Pearson Education Limited
Edinburgh Gate, Harlow,
Essex CM20 2JE, England
and Associated Companies throughout the world

© Longman Group UK Limited 1992
All rights reserved; no part of this publication may be reproduced, stored in a retrieval system, or transmitted in any form or by any means, electronic, mechanical, photocopying, recording, or otherwise, without the prior written permission of the Publishers.

First published 1992
20 19 18 17
IMP 30 29 28 27 26 25 24

Set in 10/11pt Helvetica

Printed in Malaysia, VVP

ISBN 978-0-582-65140-1

Acknowledgements

The Publishers are grateful to the following for their permission to reproduce photographs:

Barnaby's Picture Library for page 74 bottom (right); Stuart Bland for pages 1, 2 (left), 2 (right), 3 top (left), 3 centre (left), 3 bottom (left), 3 bottom (right) and 17 (eight photographs); J. Allan Cash Photo Library for pages 18 left (top), 18 centre (centre), 18 right (top), 18 right (centre), 43 and 74 bottom (left); Picturepoint for page 49; Stanley Tools for two photos on page 17 and Tropix Photographic Library for pages 18 left (bottom), 18 centre (top), 18 centre (bottom), 18 right (bottom), 74 top (left), 74 top (right), 74 centre (left) and 74 centre (right).

We also acknowledge the assistance of S. Bland, who gave editorial advice, contributed Chapter 2, Freehand Sketching, and interpreted and drew all the illustrations.

Preface

This is the third of three books which provide comprehensive coverage of secondary school Technical Drawing syllabuses. The other two books deal with plane and solid geometry and mechanical drawing.

The books will also be useful as a foundation to students taking courses in Engineering and Architecture.

In this book we shall deal comprehensively with the principles of building drawing.

The book can be used either as a class text or a self study work book.

Contents

1
Drawing office practice — 1

Drawing equipment — 1
Sheet layout — 4
Borderlines — 5
Title blocks — 5
Dimensioning — 6
Units in dimensioning — 6
Positioning the figures — 6
Dimension lines — 6
Extension lines — 6
Arrowheads — 6
Methods of dimensioning — 7
Aligned dimensions — 7
Symbols and conventions — 9
Exercises — 9

2
Freehand sketching — 10

Isometric projection — 10
Cabinet oblique projection — 10
Planometric projection — 11
Single-point perspective — 11
Two-point perspective — 11
Plans and elevations — 12
Sketching aids – the early stages — 12
Isometric grid — 12
Squared paper — 12
Sketching procedures — 12
Isometric — 12
Oblique — 13
Exercises — 13
Single-point perspective — 14
Two-point perspective — 14
Various shapes — 14
Exercises — 17

3
Orthographic projection — 19

First and third angle projections — 20
First angle projection — 20
Third angle projection — 21
Exercises — 21

4
Building construction drawings — 22

General principles — 23
Site plan — 23
Components of a site plan — 23
Floor plan — 25
Sectional views of buildings — 27
Room planning and specifications — 28
Exercises — 30

5
Foundations — 31

Depth of concrete in foundations — 31
Strip foundation — 31
Other types of foundation — 32
Exercises — 33

6
Floors — 34

Ground floors — 34
Upper floors — 37
Reinforced concrete floors — 38
Exercises — 39

7
Wall openings — 40

Jambs and reveals — 40
Window sills — 41
Lintels — 42
Arches — 43
Exercises — 47

8
Doors and windows — 48

Doors — 48
Fixed door frames — 48
Door sizes — 49
Classification of doors — 49
Representing doors on plans — 52
Determining the width of door openings — 53
Representing doors in elevation — 53
Drawing a door in section — 54

iii

CONTENTS

Windows	**55**
Window and frame sizes	55
Locating windows in walls	56
Drawing windows in sections	59
Exercises	60

9
Stairs **61**

Types of stairs	**61**
Straight-flight stair	61
L-stair	61
Dog-leg stair	62
Open-well or open-newel stair	62
Geometrical stair	63
Stair design	**63**
Terms used in connection with stairs	63
Drawing of stair details	64
Drawing a stair with a landing and a turn	68
Exercises	70

10
Roofs **71**

Components of a roof	71
Classification of roofs	72
Roof details in a building drawing	74
Roof plan	75
Exercises	76

11
Electrical wiring plans **77**

Electricity in buildings	77
Types of lamps used in buildings	77
Components of an electrical wiring plan	77
Electrical wiring symbols	77
Locating electrical components on a wiring plan	77
Exercises	79

12
Drainage and sanitation systems **81**

Water supply systems	**81**
Designing a plumbing system	81
Hot water supply systems	83
Design and layout of a drainage system	83
Disposal of waste from buildings	87
Exercises	89

Index **90**

1. Drawing office practice

Drawing equipment

The building draughtsman makes use of a wide variety of drawing instruments. The beginning student should purchase good quality instruments since these last longer, although they may be more expensive, than the poor quality ones. Students should study and understand the different drawing instruments available and how they function in order that they may use these instruments properly. Rough handling of drawing instruments reduces their quality and accuracy and also results in poor drawings.

Some of the common drawing tools and equipment used in practice for building drawing are described below. For more details concerning instruments and equipment see *Technical Drawing Book 1: Plane and Solid Geometry*.

Pencils

Drawing pencil leads are graded according to their degrees of hardness as follows: 9H (the hardest), 8H, 7H, 6H, 5H, 4H, 3H, 2H, H, F, HB, B, 2B, 3B, 4B, 5B, 6B and 7B (the softest).

The lead is encased in either wood or a mechanical holder or shaft.

To select the grade of pencil, first consider the type of linework required. For light construction lines, extension lines for dimensioning and for accurate geometrical constructions, use a hard pencil of the range 3H to 6H. Border lines for sheet layout and the outlines of building details should be black. The type of pencil chosen must be soft enough to produce black lines but hard enough not to smudge too easily or permit the pencil point to break under normal pressure. The pencil used for object lines varies from HB to 2H. Pencil grades B to 7B are too soft to be useful in drawing. Soft pencils are more useful in art work.

Always keep the drawing pencil sharp. Only a sharp pencil is capable of making clean-cut black lines. A dull pencil produces fuzzy, indefinite lines. The photograph shows well sharpened pencils.

Guidelines for sharpening pencils
1. Sharpen the end opposite to that on which the pencil grade is written in order to preserve the identifying grade mark.
2. Avoid sharpening a pencil over the drawing board or other materials.
3. Avoid sharpening the pencil point to look like a needle point.

Erasers

Erasers are available in many degrees of hardness and abrasiveness. When erasing pencil work, especially during the construction stage of a drawing, use a soft pencil eraser. Do not use gritty hard erasers, even for erasing inkwork, as they usually damage the paper.

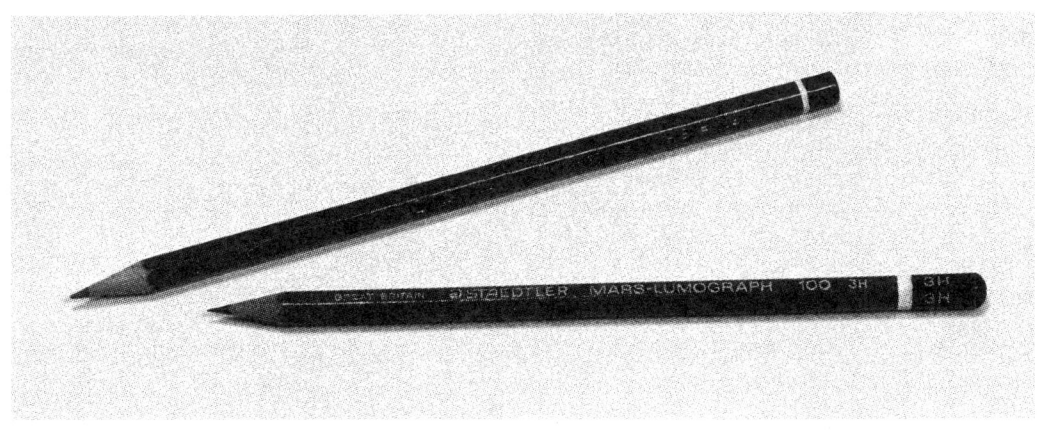

1

TECHNICAL DRAWING 3: BUILDING DRAWING

An erasing shield may be used to mask the parts of the drawing to be retained. A shield consists of a piece of metal or plastic with perforations of various shapes.

T-squares
The T-square is used to draw horizontal lines and provides an edge for guiding set squares and lettering stencils.

The upper edge of the blade is the working edge and must be straight. Some T-squares, especially those made from metals, plastics or combinations of plastic and wood, have two parallel working edges. Other T-squares have only one working edge, with the lower edge of the blade usually tapered.

A transparent T-square blade is preferred since it permits the draftsman to see the drawing in the vicinity of the lines being drawn.

Never cut paper along the working edge of the T-square as the wood or plastic can easily be cut. Even a slight nick will ruin the T-square.

Set squares
Set squares are used in conjunction with the T-square to draw vertical lines as well as angles of 45°, 30°, 60° or a combination of these. They are usually made of plastic, which has replaced metal and wood because of its transparency. Adjustable set squares (see photograph) can be used to draw lines at virtually any angle. Take care to adjust the set square to the correct angle.

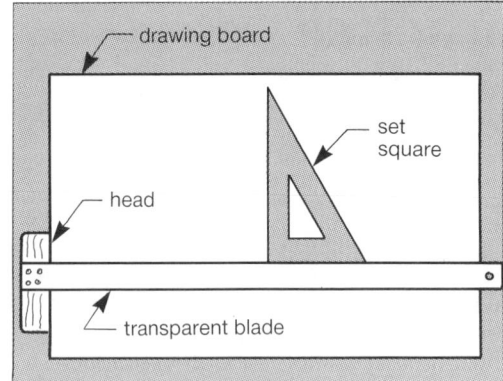

T-square and set square

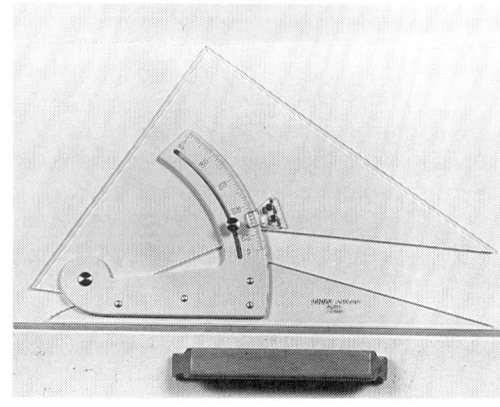

Adjustable set square on a T-square

Scale rules
Scale rules are used for plotting dimensions prior to drawing and for scaling from finished drawings. Most scale rules are made from PVC, which is dimensionally stable. The photograph shows a scale rule used by architects. The scale of 1:10 560 means that every 1 mm marked on the drawing represents 10 560 mm on the building. Different scales are used depending on the amount of enlargement or reduction required.

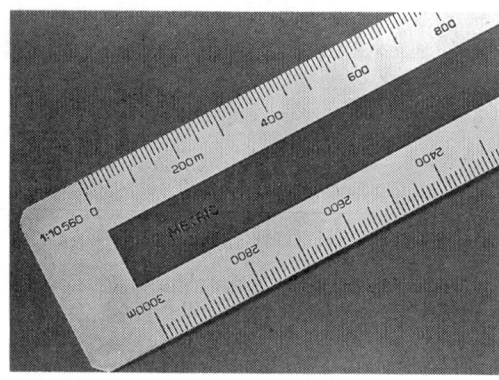

Scale rule

A very popular and useful scale of oval section with scales on four edges has the following scales: 1:10 and 1:100; 1:20 and 1:200; 1:5 and 1:50; 1:1250 and 1:2500.

Compasses
Compasses are used for drawing circles and curves. They are usually 125 mm or 150 mm long. Compasses with lengthening bars for drawing large-radius curves are also available. Compasses are designed to take a pencil lead in one arm. The pinpoint and the pencil point of the compasses should be self-centring so that even pressure can be maintained on both points. The pencil should be sharpened to a chisel point in order that thin lines may be obtained. Care should be taken in handling compasses to avoid damage to the needle point.

DRAWING OFFICE PRACTICE

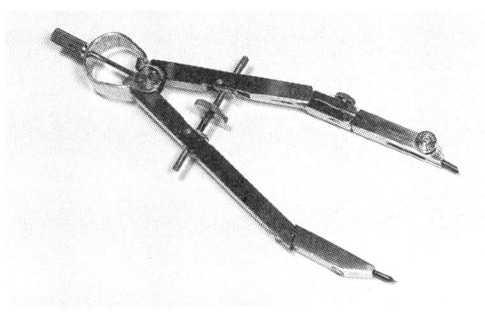

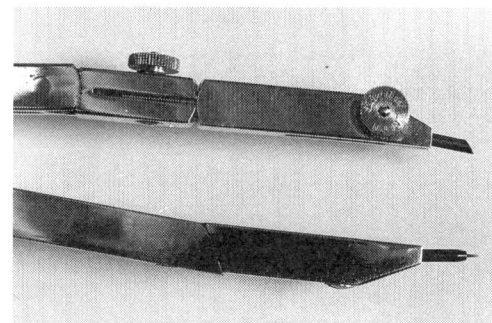

Close-up of compasses showing pencil lead sharpened to a chisel point

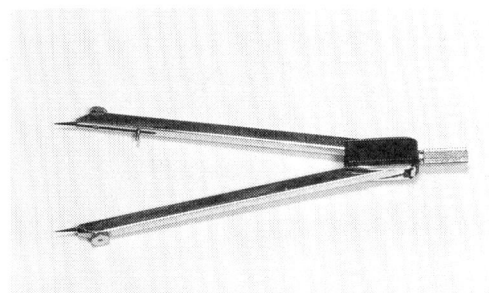

Pair of dividers

Dividers

Dividers are used to divide given straight lines or curved distances into any number of equal parts. They are also used to transfer measurements, and to set off a series of equal distances by trial and error. Some dividers are equipped with a hair-spring adjustment for small variations in the spacing of the divider points, and if the instrument is held correctly, all adjustments can be made with one hand.

The compasses can perform the same functions as a divider. Often, the compass is preferred for measuring and laying off distances, thereby reducing the number of prickholes in the drawing paper.

Drawing pens

These sometimes come under such brand names as 'Rapidograph', 'Graphos' etc., depending on the make. They are essentially fountain pens that make use of Indian ink and have interchangeable nibs for producing lines of different thicknesses. A set of drawing pens contains nibs that range in size from 0.1 to 0.8.

These pens are very useful in tracing and can also be used in lettering and inking of drawings.

Drawing pens are sensitive instruments and the manufacturers' instructions must be strictly followed. Do not try to pull out the thin wire in the nib.

Lettering guides

There are many lettering aids available to the draftsman. One such aid is the stencil or lettering guide. This consists of the alphabet in the capital and lower-case forms, numerals as well as other figures. By using the lettering guide it is possible to achieve absolute uniformity in lettering, but they are slower to use than freehand and they make the lettering look mechanical.

When using stencils do not allow the edge to rub over previous work until it is dry, otherwise the ink will spread over the drawing. Lettering stencils are available in different sizes, corresponding to the different drawing pen sizes. The letters and figures may either be upright or slanting.

Drawing pen

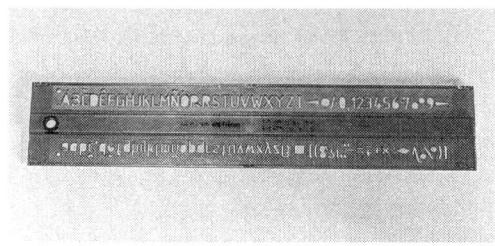

Plastic lettering stencil

3

TECHNICAL DRAWING 3: BUILDING DRAWING

Sheet layout

Drawings should be arranged logically and neatly on a sheet to give a balanced layout. Proper layout gives the drawing an agreeable appearance and makes for easy reading of the drawing. When a drawing is crowded on one side or corner of the drawing sheet it leaves the impression that there is still more work to be done.

Factors that influence sheet layout include:

1. *Need for balance*: The drawing or drawings should appear balanced on the sheet without unnecessary overcrowding of any part of the sheet.
2. *Symmetry*: Drawings should be placed centrally on the sheet. Compare a) and b) and note the improved appearance of b).
3. *Size of paper*: Whenever possible, choose a scale appropriate to the size of the drawing sheet. If the scale and size of the drawing sheet are specified, do not alter the scale but balance and symmetry must be considered.
4. *Number of drawings to be made*: If only one drawing is to be made, then it should be positioned centrally on the drawing sheet. In the case of multiview drawings or in an examination situation where the student may be required to put many different drawings on one sheet, good judgement should dictate how to locate the different views or drawings. An example of good sheet layout is shown on the right.

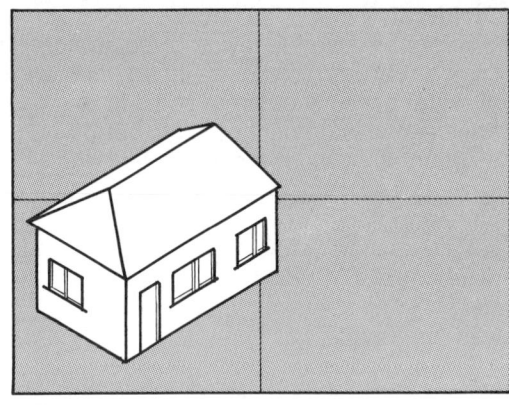
a) Badly positioned

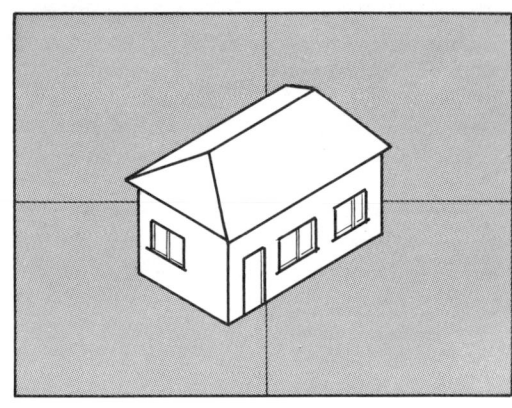
b) Correctly positioned

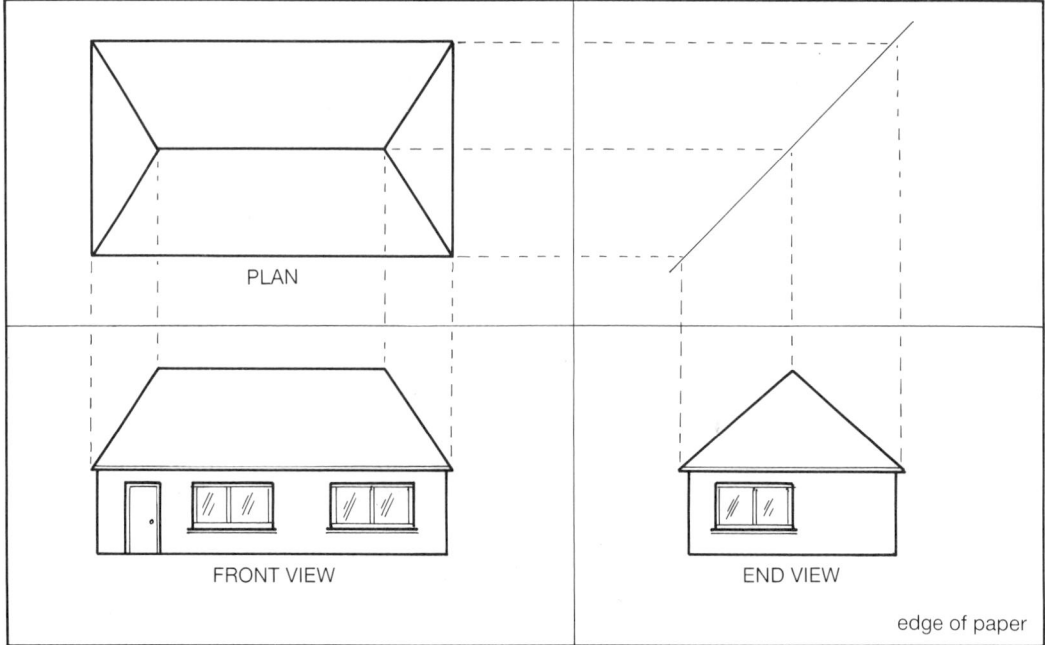
Sheet layout in orthographic projection

DRAWING OFFICE PRACTICE

Borderlines (or margins)

Drawing practice requires that every drawing sheet shall have a margin or borderline. It defines the area which should contain the drawing. The margin should normally be between 13 mm and 20 mm from the edge of the drawing sheet. However, under examination conditions, it is important to draw margins as may be specified by the examiner.

Title blocks

The title block or information box is placed at the bottom corner or across the length of the sheet at the bottom.

Corner title block

In this method the title block is located at the bottom right-hand corner of the drawing sheet. The information supplied includes the job title, subject of drawing, name of student, name of school or examination number (where applicable), scale and date of drawing. Sometimes the names and initials of the persons who drew, traced, checked and approved the drawings are added.

Under examination conditions students are not required to provide all necessary information, but examiners and draftsmen generally agree that information regarding the name of the person who made the drawing as well as the subject of the drawing or title, scale, school or address, date, registration number or examination number are important. An example is shown on the right above.

A box with height measuring 20 mm to 50 mm is usually sufficient.

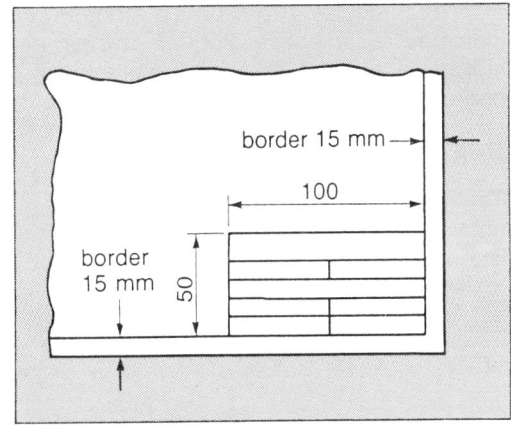

Position of title block

Corner title block

Full-length title block

In the full-length method (see below) the title block extends the entire length of the bottom of the drawing sheet. Consequently, the height is often less than the 50 mm used for the corner block. The length of the box may be divided into three equal parts depending on the size of the sheet and the amount of information to be supplied.

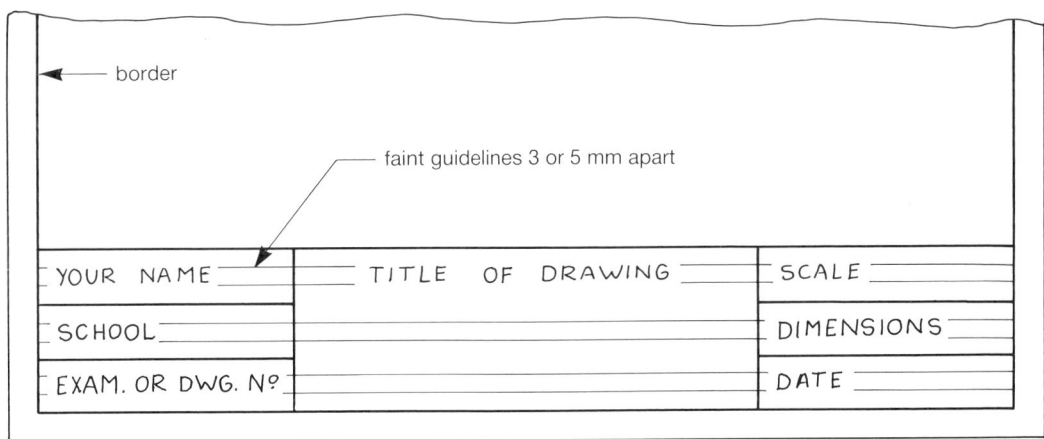

Full-length title block

Dimensioning

For a building to be constructed according to design specifications, the working drawings must be correctly and accurately dimensioned. Dimensions are used on a drawing to indicate the size and relative location of the building details. The placement of dimensions requires good judgement, since any misinterpretation of a dimension could result in a major construction error.

When dimensioning a drawing, the following rules should be remembered.
1. *Accuracy*: The values indicated on dimensions must be accurate to the same number of decimal places.
2. *Clarity*: Each dimension must be placed in its most appropriate position, that is, where it should be seen easily and clearly.
3. *Completeness*: There must be no omissions of specifications.
4. *Readability*: Lettering, numerals and dimension lines must be neat, uniform in size, and very distinct.

Units in dimensioning

Dimensions in building drawings are usually given in millimetres (mm). However, drawings may be dimensioned in metres (m), especially when the figures are relatively large, in which case the figure is corrected to three places of decimals (see above right).

It is bad drawing practice to use the millimetre and the metre in one drawing. When either metres or millimetres are used it is not necessary to write m or mm after each figure. The distance alone is indicated: for example, 2500. However, an indication must be made in the title block of the unit used.

Positioning the figures

The dimension figure may be placed above the dimension line or the dimension line may be broken and the figure written in the middle.

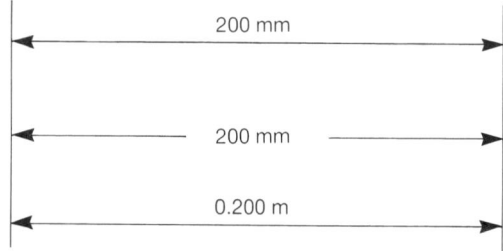

Alternative dimensioning methods

Dimension lines

Dimension lines are used to indicate the limits of a detail. A dimension line is usually a thin, dark line which is either continuous or broken in the middle and which indicates the direction and extent of a distance. When drawings are crowded with dimensions it is difficult to see the main object, therefore the dimension lines should be at least 20 mm from the drawing outline and at least 6 mm apart.

Extension lines

These are thin continuous lines extending from, but not touching, the point to which the dimension refers. The extension line

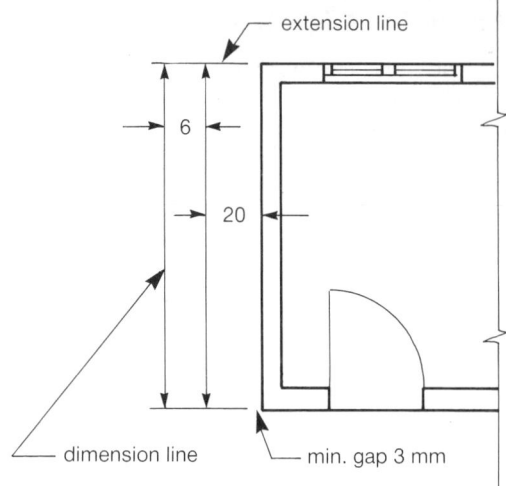

Spacing dimension lines

continues a short distance on either side of the arrowheads of the dimension line. It should be at least 3 mm from the object line. For internal dimensions, for example rooms in a building, the internal walls of the room serve as extension lines (see above).

Arrowheads

Arrowheads are used to represent the ends of dimension lines. There are several ways of drawing arrowheads and those shown on page 7 are generally accepted in building drawings. Only one method should be used in a given drawing. The first method is the most difficult to draw, although commonly used. The student should, therefore, learn to draw well formed and sharp arrowheads.

DRAWING OFFICE PRACTICE

Methods of dimensioning

There are two common methods of dimensioning building drawings. In method 1 (see below) the dimensions indicated represent the space between the two immediate arrowheads. This is the more popular method and is easier to interpret. In method 2 the dimensions are read from one point known as the datum and all the arrowheads run in one direction. The datum is usually represented with a dot. Distances are added to each other, beginning from the datum.

Aligned dimensions

The aligned system is often adopted in building drawing and in drawings of civil engineering projects. All dimension figures are aligned with the dimension lines so that they may be read from the bottom or from the righthand side of the sheet. The isometric drawing below shows a block dimensioned in the aligned system. Note that the dimension figures are always parallel to the dimension lines and never perpendicular to them.

Arrowheads for building drawings

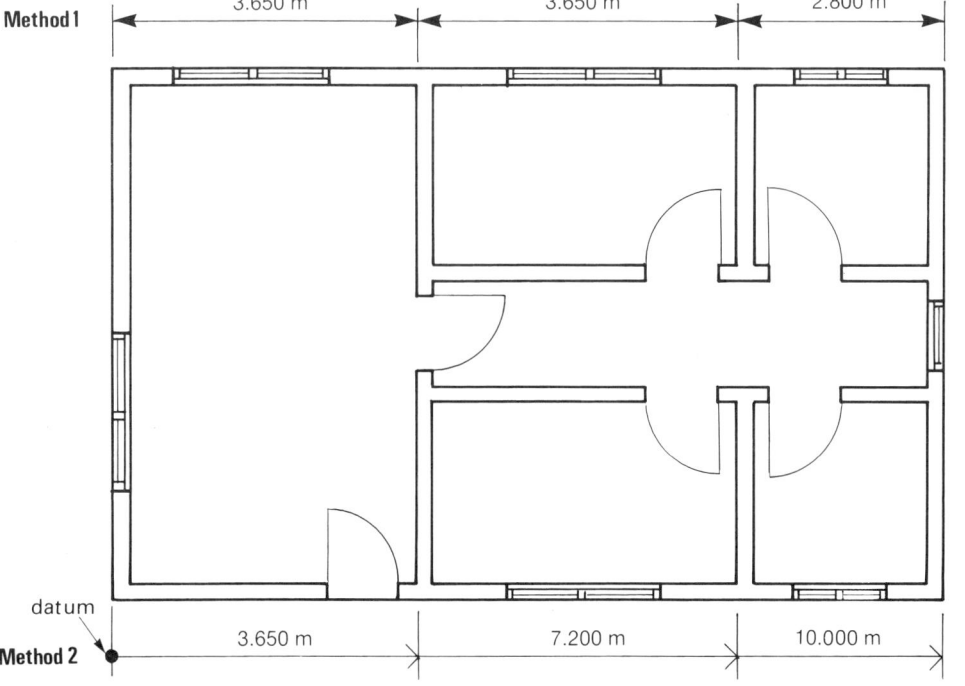
Methods of dimensioning

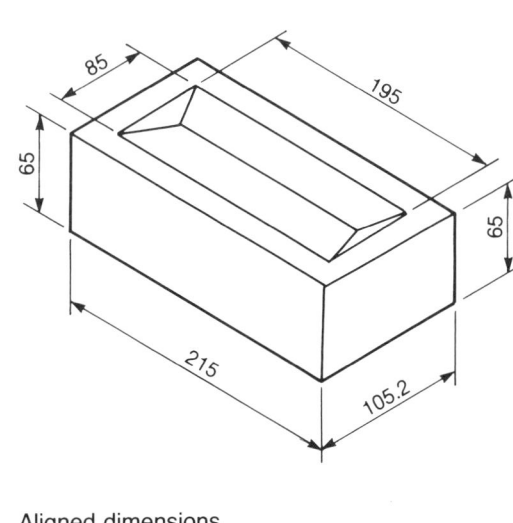
Aligned dimensions

Sometimes drawings may be so crowded that there are insufficient spaces for all the information required in dimensioning. The information left out must be included in the title block.

7

TECHNICAL DRAWING 3: BUILDING DRAWING

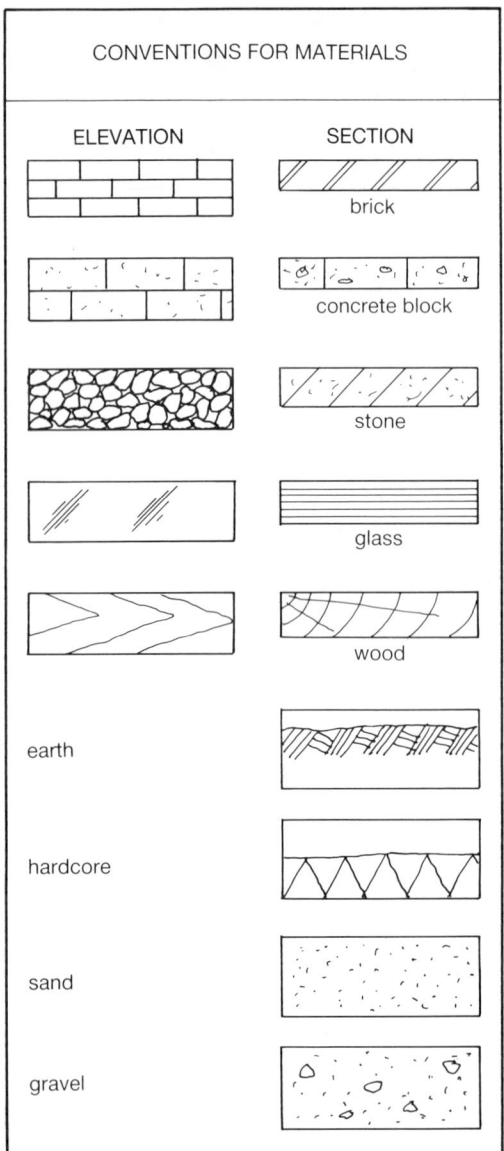

CONVENTIONS FOR MATERIALS	
ELEVATION	SECTION
brick	brick
	concrete block
	stone
	glass
	wood
earth	
hardcore	
sand	
gravel	

SYMBOLS ON SITE PLANS	
north point	bench marks
	BM X 752·09
	BM △ 921·65
trees	property corners
	EL 39·7
	EL 63·4
cultivated area	
woodland	
large stones	
property line	— — — —
fence	—X——X—
railway	+++++++++++
paved road	
unpaved road	- - - - - -

ABBREVIATIONS	
aggregate	Agg
air brick	AB
asbestos	abs
boarding	bdg
bench mark	BM
building	bldg
brickwork	bwk
damp proof course	DPC
damp proof membrane	DPM
finished floor level	FFL
foundation	fdn
ground level	GL
hardcore	hc
joist	jst
reinforced concrete	RC
tongue and groove	T&G

8

Symbols and conventions

Symbols and conventions are used to save time and avoid confusion which would otherwise arise in highly detailed drawings and in interpreting the drawing of materials that look alike.

A symbol is a sign or mark used to represent an object, idea or process. A convention is an accepted standard which has been adopted for clarity (although it may violate true orthographic projection).

Materials in section on plans and sectional elevations are usually shown hatched (diagonally shaded). Hatching is preferable to other methods such as colouring for distinguishing different materials because it is less laborious, less costly and less liable to errors.

The symbols used in elevations and sections for various building materials are shown on page 8.

Exercises

1. Select two pencils, one 'H' and one 'B', and sharpen these ready for use in drawing.
2. Draw several lines on your drawing paper. Then erase these lines:
   a) using eraser shields,
   b) without using eraser shields.
3. Set up your drawing paper on a drawing board. Then try to draw straight lines on your sheet:
   a) using a T-square only,
   b) using a T-square and set squares,
   c) without the use of instruments.
4. Using your scale (see Book 1, page 11) or scale rule, reduce each of the lengths 100 m, 25 m and 45 m to each of the scales 1:10, 1:100 and 1:500. Do you notice any differences in the lengths given? Draw these lengths using a scale of 1:100. Measure the length (to scale) of each line.
5. Draw the following, using your compasses:
   a) a semicircle, radius 10 mm,
   b) an arc, radius 25 mm,
   c) a circle, radius 48 mm.
6. Using your pair of dividers, divide the following lengths into the number of parts indicated:
   a) 150 mm into 6 equal parts,
   b) 288 mm into 6 equal parts,
   c) 105 mm into 7 equal parts,
   d) draw any line and divide it into any number of equal parts of your choice.
7. Prepare a full-length title block. In it print the following information
   a) without a lettering guide,
   b) using a lettering guide if available: your name; your school – DRAWING NUMBER 1; the title of your drawing – 'EXERCISES'; scale – 1:1 (FULL SIZE); dimensions in mm; the date.
8. Lay out your drawing paper on a drawing board and on it insert the following at their appropriate places:
   a) borderlines,
   b) title block.
9. Draw each type of line indicated below:
   a) object line,
   b) centre line,
   c) section line.
   Try to draw the other types of lines used in technical drawing (see Book 2, page 2).
10. Show by means of appropriate sketches the symbols or conventions for these materials:
    a) concrete,
    b) wood, in section,
    c) screed,
    d) rock,
    e) earth,
    f) hardcore.
11. Using isometric projection, draw a box 2000 mm × 1550 mm × 900 mm to a scale of 1:10. Use aligned dimensions to dimension the box. Arrowheads, extension lines and dimension lines should be correctly shown.

2. Freehand sketching

Freehand drawings are made when information and ideas need to be quickly put on paper. Working drawings are made using instruments, accurately and to a scale. The working drawing is often a development from freehand work which is done at the early planning and design stage.

A sketch can quickly give an overall impression of the shape and proportions of an object. Various forms of three-dimensional presentation are shown below.

The following rules apply to freehand work:
1. Drawings should be neat, outlines should be continuous and sharp, and construction lines should be faint. A softer pencil is usually used, H or HB.
2. The drawing should be in proportion, i.e. the shape of the object should not be lost. The sketch should give a true impression of the object.
3. The edges of rules and set squares must not be used. You should estimate lengths and angles, and lines should be drawn without the use of instruments.

Isometric projection

Isometric projection is the most common method for sketching tools and machine parts. The sides are angled at 30°. All sides are distorted in shape.

Cabinet oblique projection

Cabinet oblique projection shows the true shape of the front of the object. The sides are angled at 45° and their lengths are halved to make the proportions of the drawing look correct.

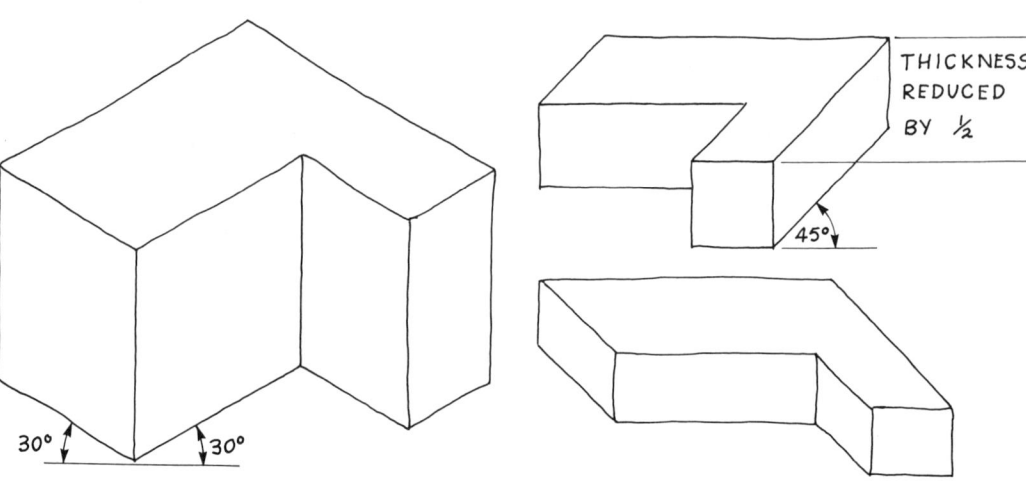

The drawing above shows more detail of the corner than the first drawing with the sides facing the right. You should position your sketches to show as much detail as possible.

FREEHAND SKETCHING

Planometric projection

Planometric projection is used to show the interiors of rooms in buildings. The true shape of the plan is drawn but it is angled at 30°, 60° as shown, or it can be angled at 45°.

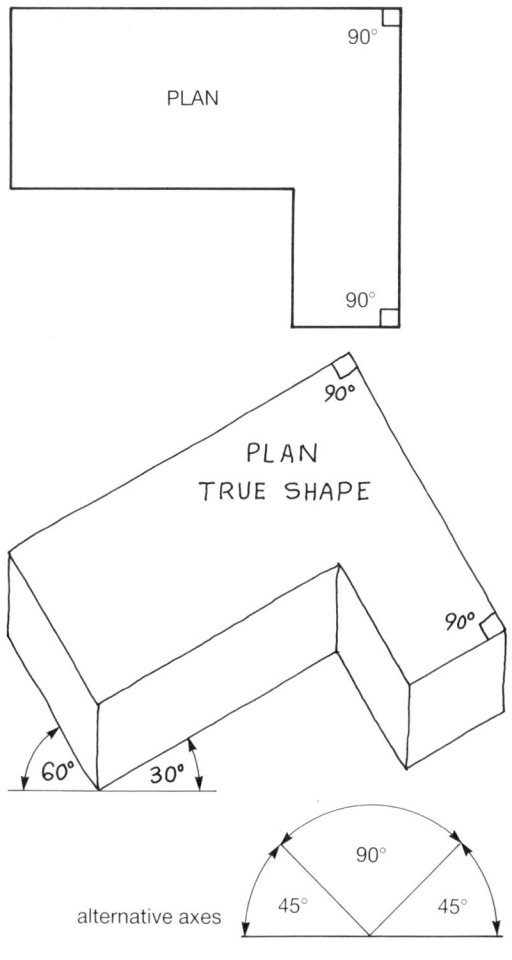

Single-point perspective

This is similar to oblique projection. The front is seen as the true shape. One vanishing point is used and its best position is above and to the side of the drawing to enable the top and the sides to be seen.

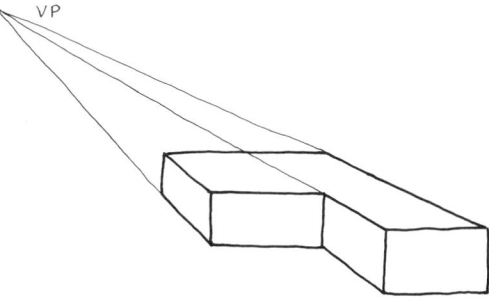

Lines radiate from the vanishing point, and the lengths of the sides are adjusted to give an impression of the correct proportions.

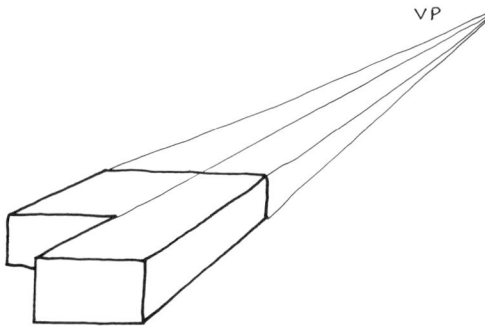

In the drawing above the vanishing point has been moved to the righthand side and the drawing shows less detail. We normally position the drawing such that the maximum amount of detail can be seen.

Two-point perspective

This is similar to isometric projection except that the sides converge to two vanishing points. With the vanishing points above the drawing the top is seen.

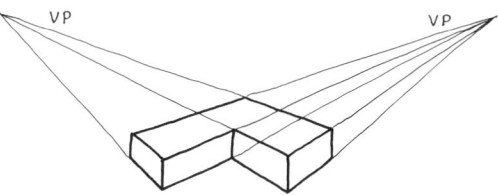

The vanishing points are located on the horizon and in this case they are level with the object. They are badly positioned in the drawing below because no detail of the top or bottom is shown.

With the vanishing points below the object the bottom is seen. For two-point perspective both vanishing points lie on the same horizontal line. The drawing is usually arranged slightly off-centre.

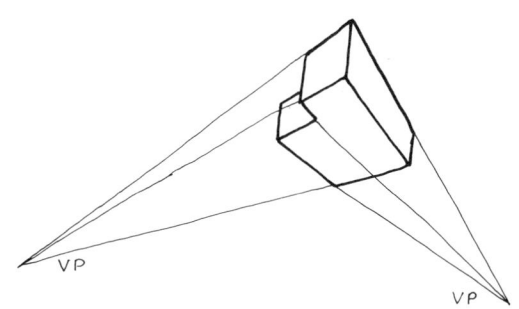

11

TECHNICAL DRAWING 3: BUILDING DRAWING

Plans and elevations

Sketching is often useful when planning the layout of buildings. Ideas can be quickly put down on paper to work out the best possible layout. Two different designs for a kitchen are shown below.

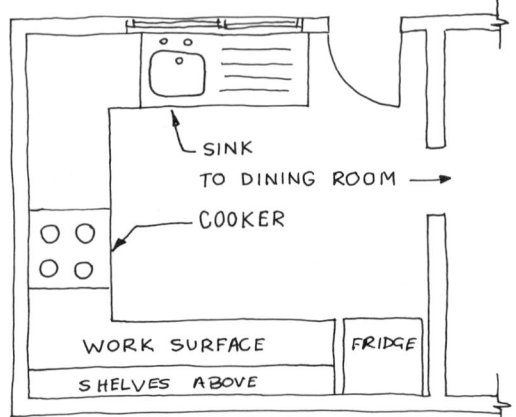

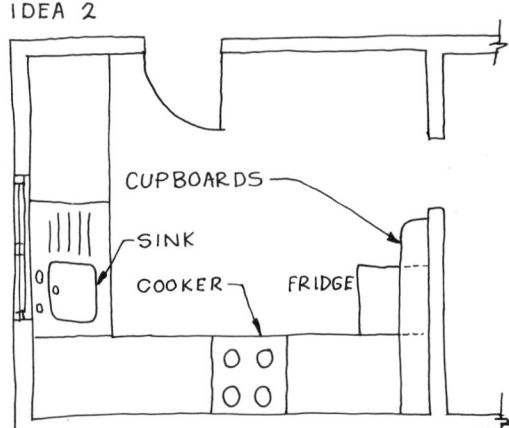

Sketching aids – the early stages
Isometric grid

Practice is needed in drawing parallel and straight lines at the correct angle for isometric projection. The drawing below shows a block sketched on an isometric grid. The lines are normally spaced at 10 mm intervals.

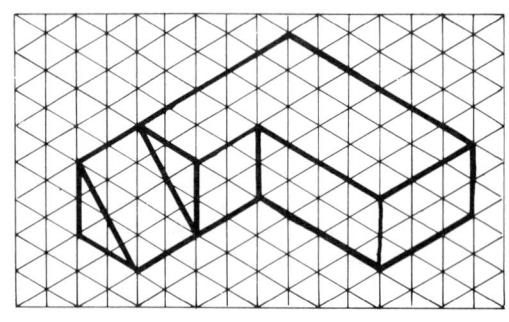

Squared paper

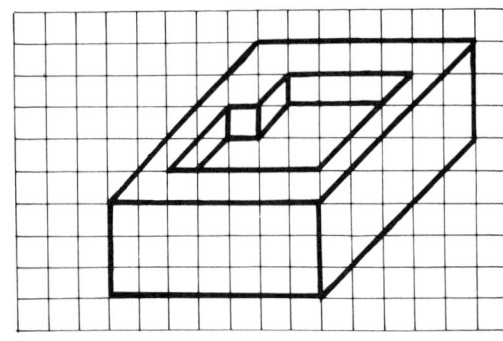

Squared paper is useful for oblique and orthographic sketches.

Sketching procedures

Isometric

The following shows the procedure for making an isometric sketch without the aid of a grid. You should estimate angles and lengths.
1. Draw the vertical and two lines at 30°.
2. Mark the height and sketch in the sides and the top.
3. Draw in the corner.
4. Complete the corner and outline the drawing. If construction lines are faint enough they should not be erased.

1.
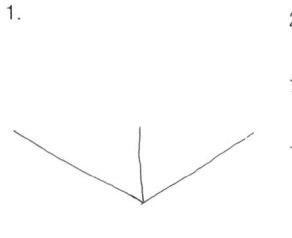

2.
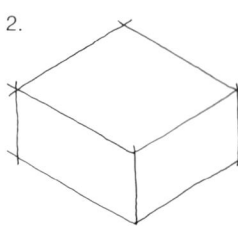

3.
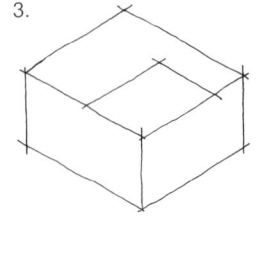

4.
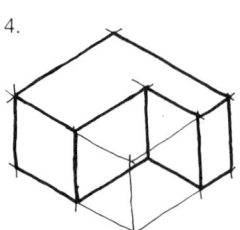

FREEHAND SKETCHING

Oblique

The following shows the procedure for making an oblique sketch without the aid of a grid. You should estimate angles and lengths.
1. The easiest way is to draw a box and then remove the corner. First draw the front of the box.
2. Draw parallel lines at approximately 45°.
3. Complete the top and draw in the cut-out corner.
4. Complete the corner and outline the drawing. Construction lines should be sufficiently faint to be left on the drawing.

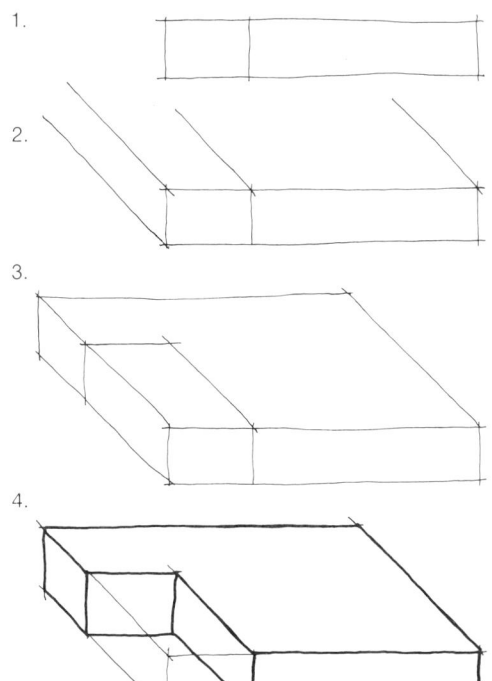

Exercises

1. Make isometric, oblique and orthographic sketches of the objects shown:

a) using grids (grids can be traced from page 155 in Book 1).
b) without grids.
2. Make perspective sketches of each object (see page 14).

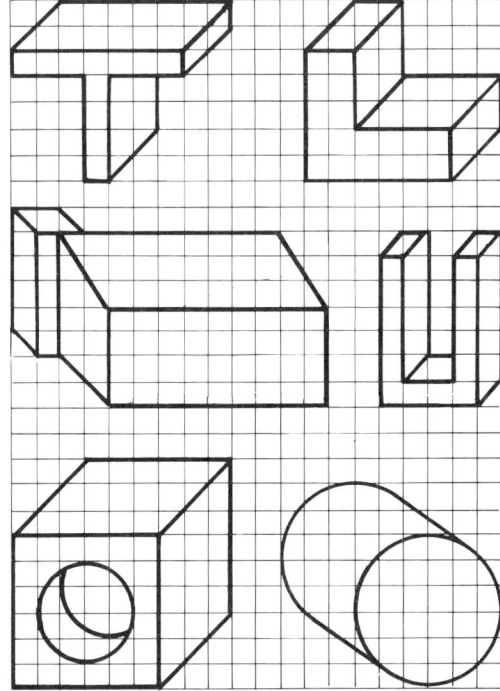

3. Can you spot the mistakes on this drawing?

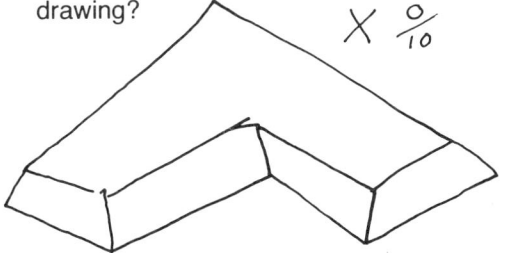

13

TECHNICAL DRAWING 3: BUILDING DRAWING

Single-point perspective

The following shows the procedure for making a single-vanishing-point (VP) perspective sketch. Arrange for the vanishing point to be above and to one side of the drawing.

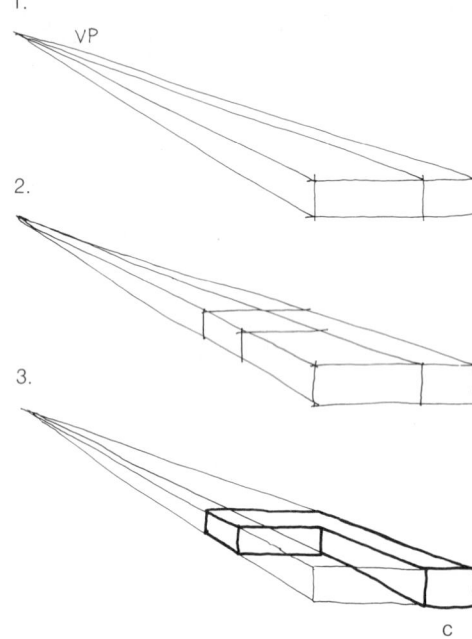

1. Sketch the front of the object to its true shape. Treat the object as a box at this stage and cut the corner out later. Draw lines to the vanishing point (VP).
2. Draw in the top and the sides. The thickness will be slightly reduced to give correct proportions.
3. From C draw a line to the vanishing point and complete the corner. Outline the sketch.

Two-point perspective

The following shows the procedure for making a two-vanishing-point perspective sketch. The vanishing points should be arranged to be above the drawing and on the same horizontal line.

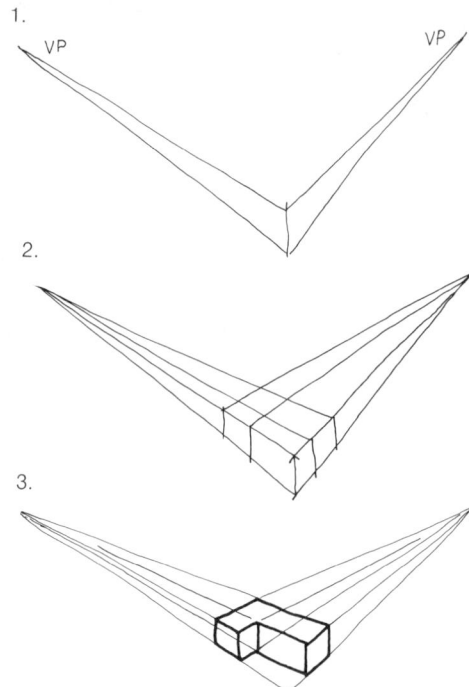

1. Sketch the front vertical corner of the box off-centre and below the vanishing points. Project lines to vanishing points.
2. Draw in sides and top of box by projecting lines to the vanishing points.
3. Locate the cut-out corner and project lines back to the vanishing points. Outline the drawing.

Various shapes

Circle
The drawings below show the procedure for sketching a circle in isometric projection.
1. Draw a square and the centre lines. To sketch the true shape of the circle, sketch in each quadrant separately.
2. Draw the same square using the isometric axes. Draw in the centre lines.
3. Sketch in the two curves. It is easier to remove your paper from the drawing board and to arrange it so that your hand is on the inside of the curve.
4. Complete the circle by drawing in the remaining two curves. The radii for these curves will be greater than that for the first two curves.

Outline the circle. Your construction lines should remain on the drawing and should be faint.

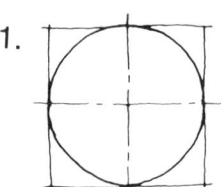 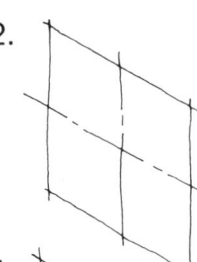
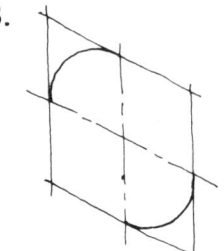 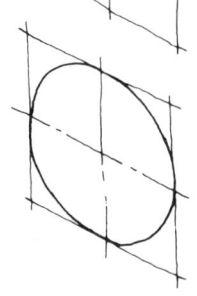

FREEHAND SKETCHING

Cylinder

The drawings below show the procedure for drawing a cylinder in isometric projection.
1. Sketch a box and add the centre lines. The two ends should be squares drawn in isometric projection.
2. Sketch the circles at both ends.
3. Sketch in the sides so that they are tangential to the circles. Outline the drawing.

This cylinder has been drawn such that it points towards the right. The same procedure applies to a cylinder pointing left or a vertical cylinder.

The same procedure can be used for sketches in oblique and perspective.

Planometric

The drawings show the procedure for making a planometric sketch of a kitchen. The walls at the front have been removed to show more detail.

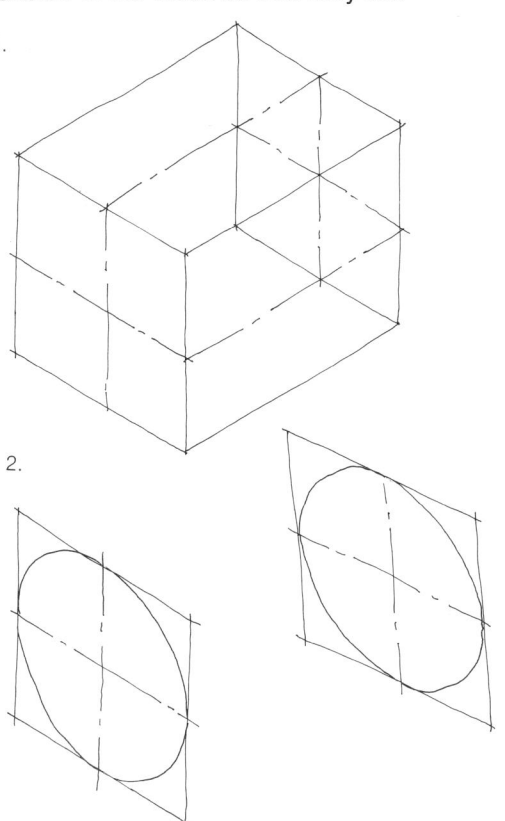

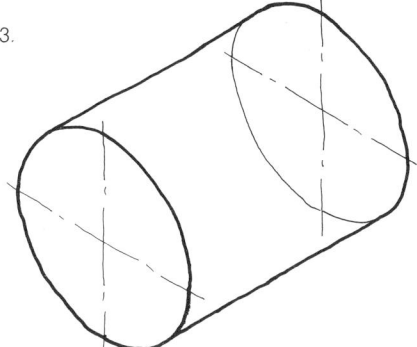

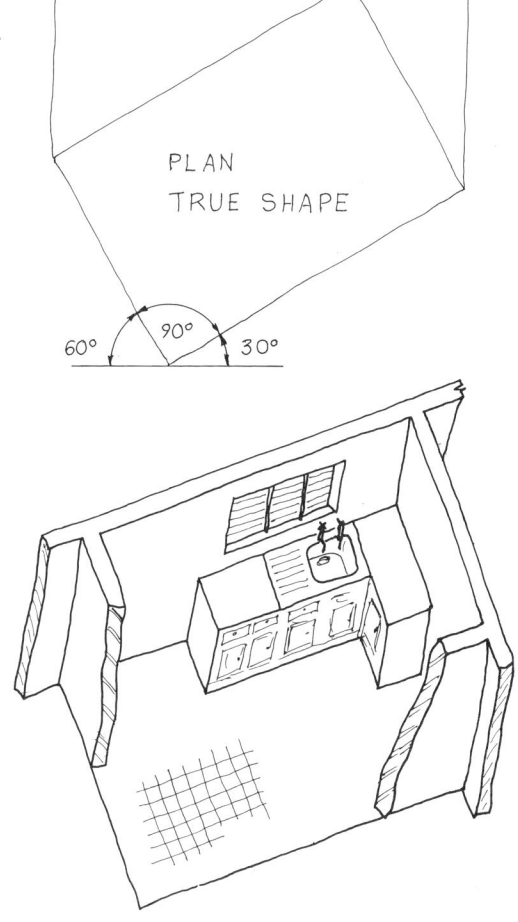

TECHNICAL DRAWING 3: BUILDING DRAWING

Perspective
The drawings below show the procedure for making a two-point perspective sketch of a small bungalow.

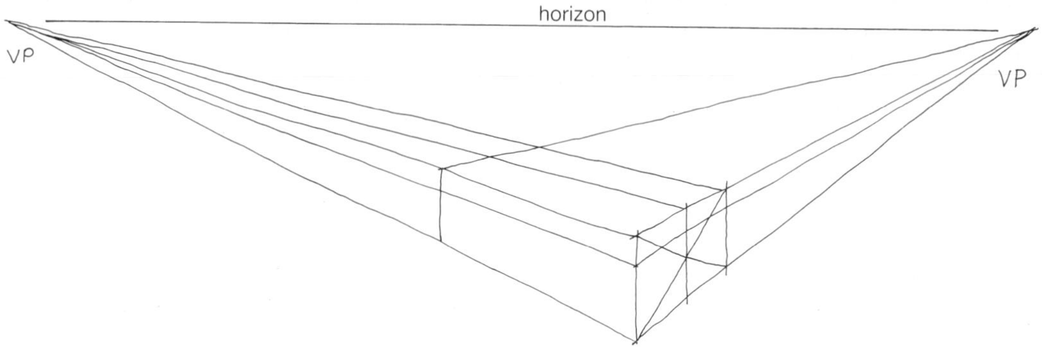

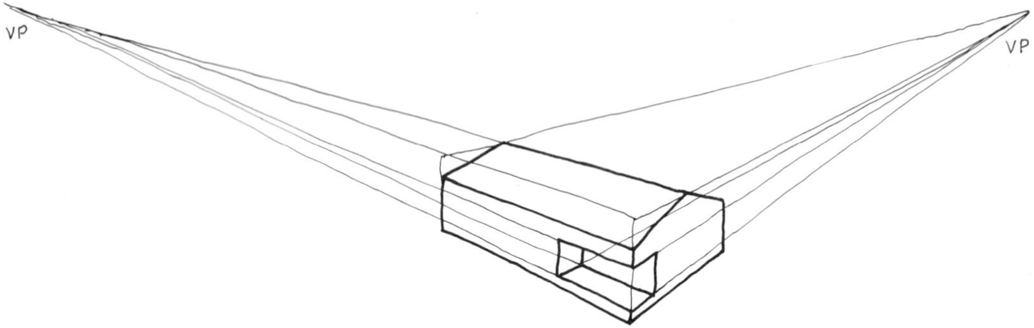

Bricklayer's trowel
The drawings show the stages for sketching the bricklayer's trowel in the photograph.

1. Draw the centre lines and boxes for the handle and blade.

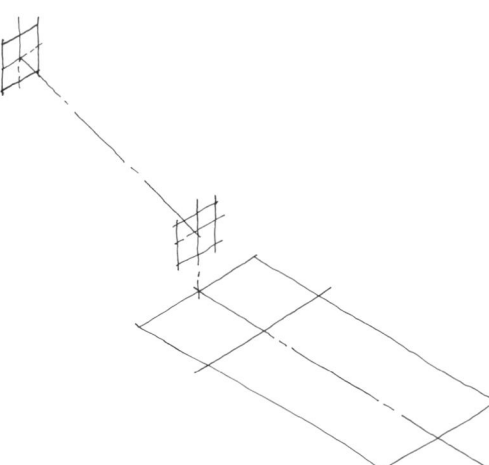

16

FREEHAND SKETCHING

2. Sketch the circles for the handle and complete the handle.

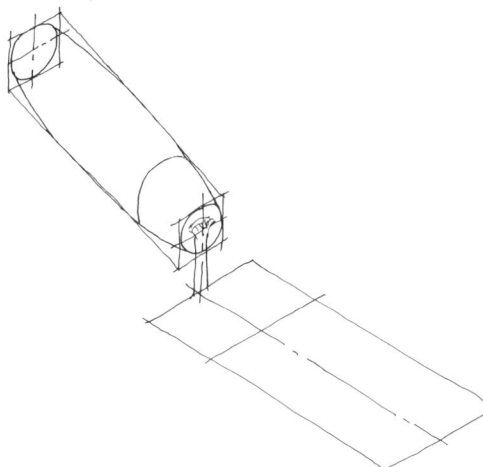

3. Draw the blade and outline the drawing. Light construction lines can remain and shading will help to make the sketch more realistic.

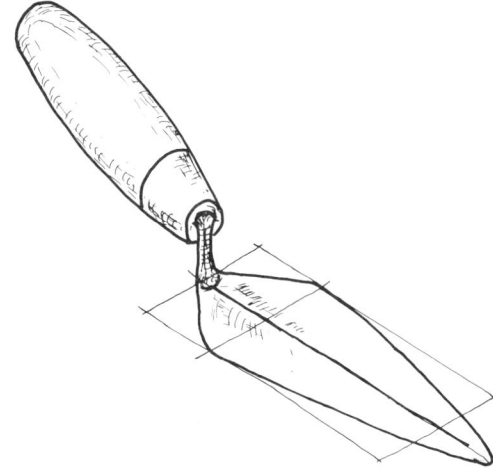

Exercises

1. Tools

The photographs show various workshop and builders' tools. Try to find an example of each tool and make freehand (isometric and oblique) sketches for each tool. Sketch the plans and elevations for some tools.

Try to find examples of other tools not shown and make similar sketches.

Your sketches should be neat and in good proportion. Position your sketch to show the greatest amount of detail.

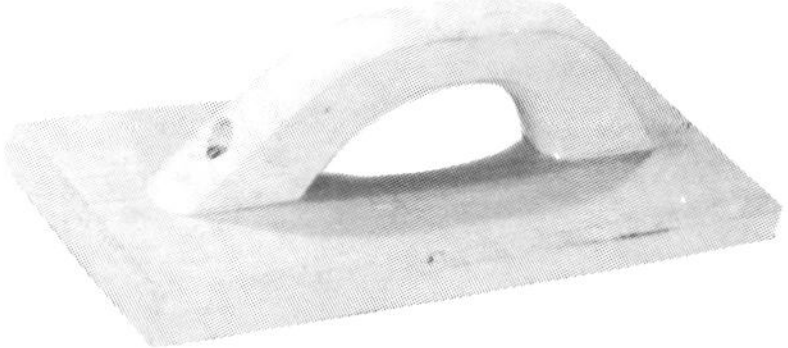

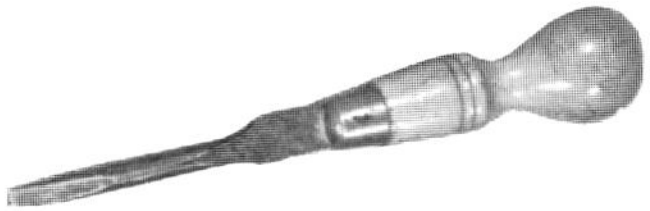

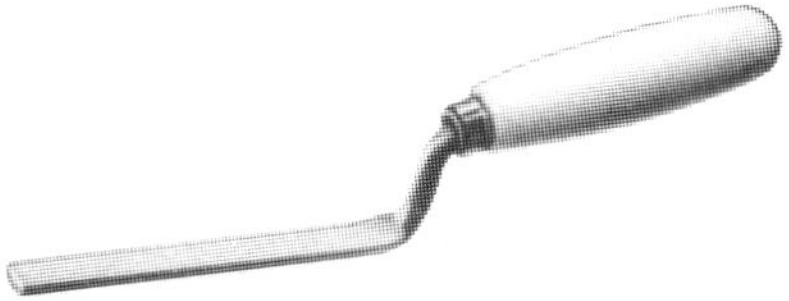

17

TECHNICAL DRAWING 3: BUILDING DRAWING

2. Buildings

The photographs show various buildings.

Try to find examples of buildings similar to the ones shown in the photographs and make perspective sketches of them.

Your sketches should be neat and in good proportion. You should view the building so as to show as much detail as possible. Try to vary the position for your vanishing points. The vanishing points should be as far apart as possible.

Try to find examples of other buildings not shown here and make freehand perspective sketches.

As an additional exercise you could sketch some of the tools on page 17 using either single-point perspective or two-point perspective.

3. Orthographic projection

Orthographic projection is very useful for making working drawings of buildings. Usually the required drawings include the elevations (front and end views), the plan or top view and the section through the object. The drawing below shows the front, top and end views of a one-storey building.

The view from A is known as the front view (or front elevation), and appears to the eyes as shown in a). A view from the top (B) is referred to as the plan and is visualised as shown in b). Similarly, a view from C, called the end elevation or the end view, is seen as shown in c).

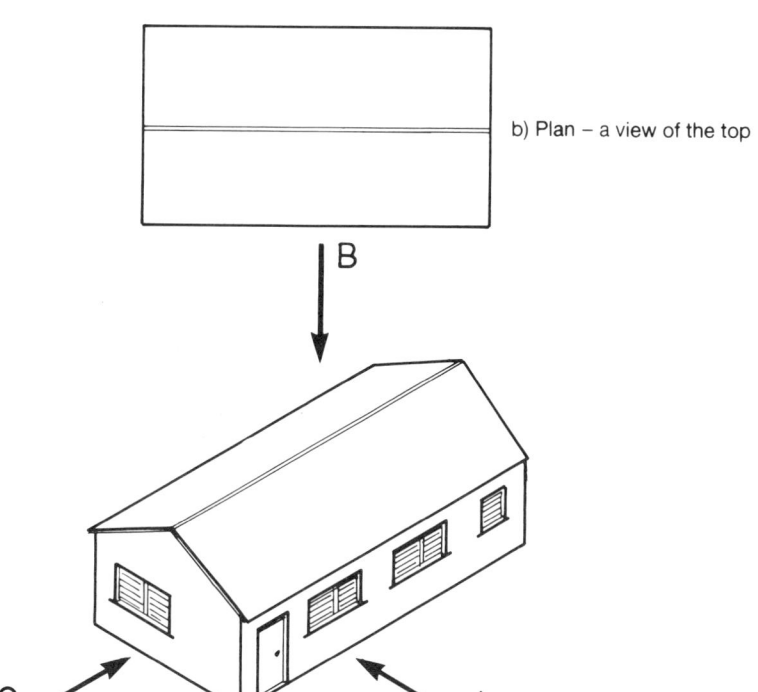

b) Plan – a view of the top

c) End elevation – a view of the end

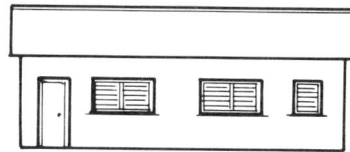

a) Front elevation – a view of the front

First and third angle projections

For orthographic projection, two planes are assumed to intersect. This line of intersection is known as the XY line or ground line and the four angles (dihedral angles) resulting from the intersection of these lines are all right angles (see drawing). The four dihedral angles are numbered for reference as 1st, 2nd, 3rd and 4th angles, and of these the 1st and 3rd angles are used in conventional practice for all projections.

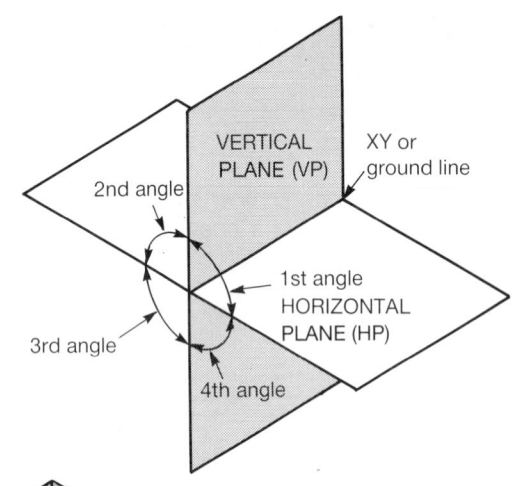

First angle projection

During projection the front view and plan are shown on VP and HP, respectively. An extra vertical plane (SVP) is used so that the end evelation can be projected on to it. Side vertical planes (SVP) can be at either or both ends, depending on the location of the detail to be shown. Where only one is required it is normally placed to the right of the front vertical plane.

One of the most important points to remember is that the lines of projection are always perpendicular to the faces of projection. The drawing below shows

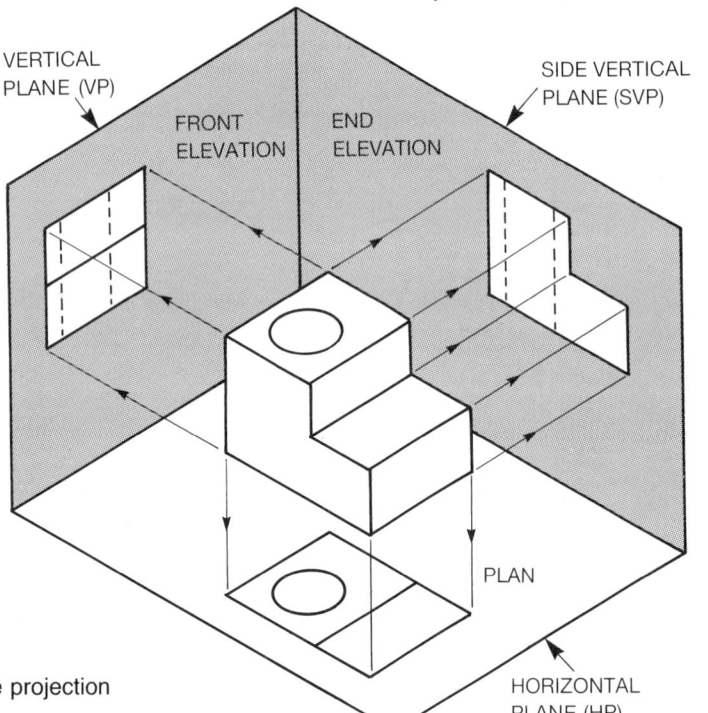

Views projected on to the planes in first angle projection

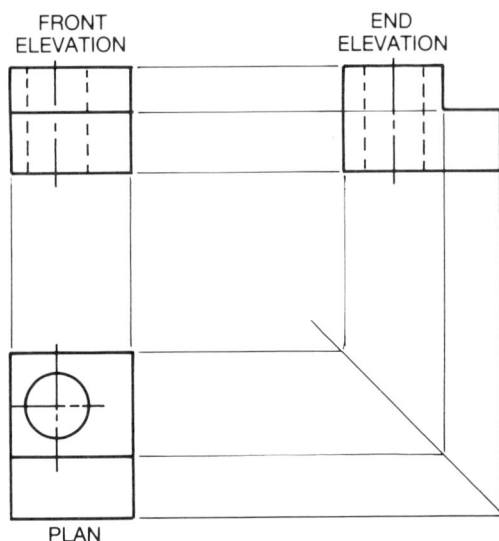

First angle projection

the projection of an object on to the three planes of projection. The planes are opened out to show the views correctly positioned in first angle projection.

The object is placed between the viewer and the plane of projection so that the view obtained from the left appears on the right of the elevation and vice versa. Similarly, the view from the top is drawn below and vice versa. This is the essential feature of 1st angle projection and one which distinguishes it from 3rd angle projection.

Third angle projection

In the 3rd angle method of projection the plan is placed above the front elevation. The simple rule is: whatever is seen on the right side is placed on the right side; whatever is seen on the left side is placed on the left. This is the reverse of the 1st angle method.

The block shown on page 20 is shown again below in 3rd angle projection for comparison.

More information on 1st and 3rd angle projection can be found in Chapter 10 of Book 1.

In practice, either the 1st angle projection or the 3rd angle is acceptable. In architectural drawing a combination of both is widely used. When using either the 1st or the 3rd angle projections a note to that effect should appear on the drawing, for example, 3rd angle projection. Alternatively, the direction in which the views are taken should be indicated.

Exercises

1. Draw to a scale of 1:100 the plan, front and end elevations of the building below.

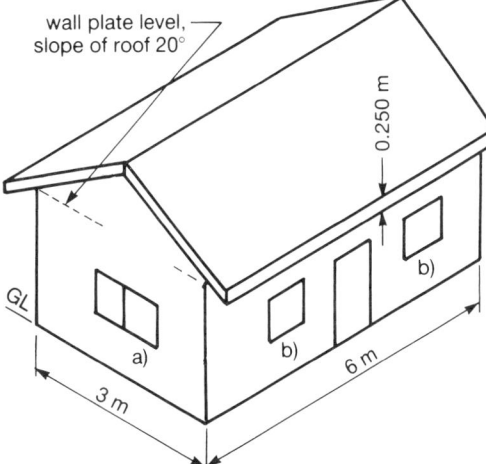

Note:
  i) The height of the wall from ground level to wall plate level is 3.000 m.
  ii) The roof 'overhangs' 0.600 m beyond the front and rear walls and 0.450 m beyond the end walls of the building.
  iii) The door is 0.750 m wide and 1.950 m high.
  iv) Window a) is 1.200 m wide and 1.000 m high.
  v) Window b) is 0.750 m wide and 1.000 m high.
2. Draw the plan, end and front elevations of the object shown below:
   a) using 1st angle projection,
   b) using 3rd angle projection.

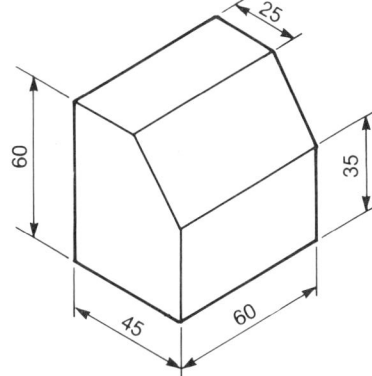

3. Look around your school compound and your community for the various types of buildings available. Sketch the elevations and plan of at least three different buildings as follows:
   a) a church building,
   b) a bungalow,
   c) a single-storey building.

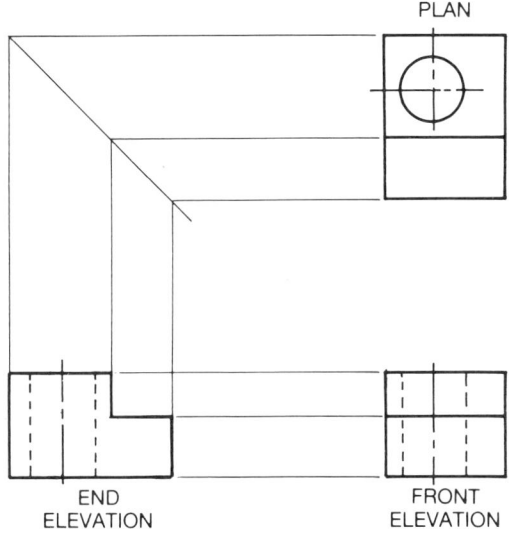

Third angle projection

4. Building construction drawings

Buildings are constructed according to projects, complete with estimates. A project is made up of all the building construction drawings required for erecting the building. Estimates specify the extent of the various kinds of jobs, their cost, and the amount of building materials, labour and equipment required for constructing a building. A good estimate also indicates the total cost of the building.

Working drawings for residential or industrial buildings are usually made on the basis of the approved project. Often, working drawings consist of the general architectural and building construction drawings including details of: the site plan, building plans, roof, floor and foundations; front, rear and end elevations; sections of foundations, stairs, floor slabs, and sections taken from the roof through the foundations. A good working drawing includes electrical and plumbing details, together with relevant explanatory notes for any of the drawings.

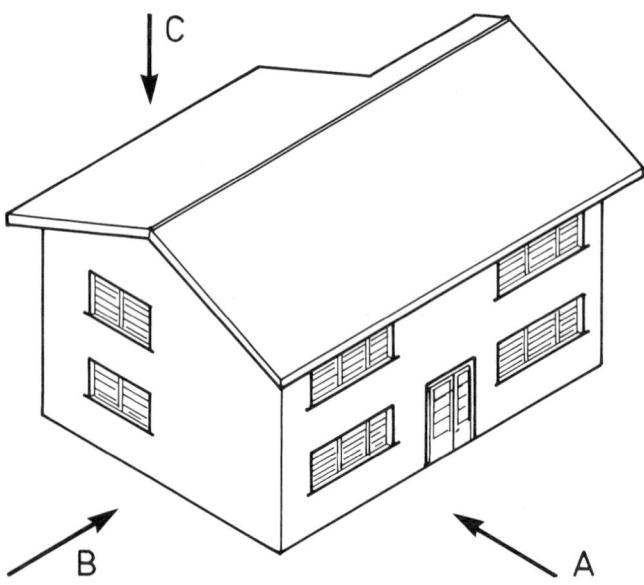

First angle orthographic views of a two storey building

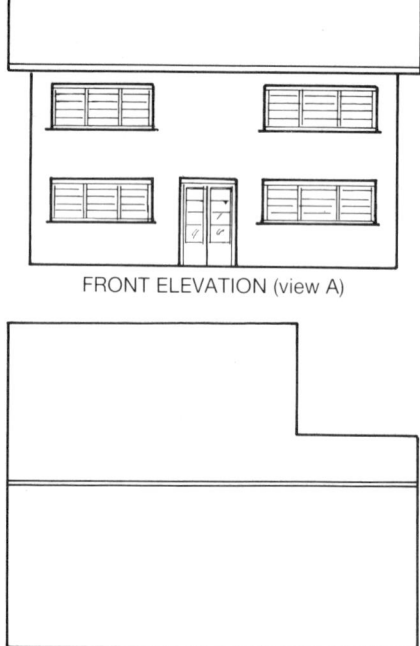

FRONT ELEVATION (view A)

ROOF PLAN (view C)

END ELEVATION (view B)

BUILDING CONSTRUCTION DRAWINGS

General principles

The principles of orthographic projection outlined in the preceding chapter apply to building drawings. The views of buildings in drawings have their own names; these are: the front elevation, which is the main view of the building taken as a person approaches the building from the street; the rear elevation; the east or west end elevation; the roof plan (see page 22). The elevations and the roof plan of a building give a graphical description of the exterior only, while the various floor plans and sections of the building show the arrangement and dimensions of the rooms in the building. The sections also show the main structural and plumbing details, sanitary appliances, and so on.

Site plan

A site is a parcel of land which is made up of one, two or more plots. A site plan, therefore, is a drawing showing various properties in terms of their owners, locations, elevations, states of development and features such as roads, utility supply lines, etc. Several components are used to represent a site plan as shown below.

Components of a site plan

These include survey beacons, contour lines, orientation symbols, physical features, access roads and utilities.

Survey beacons

These are concrete pillars, about 750 mm long, which are buried in the ground leaving about 75 mm above ground level. Survey beacons are located at principal corners of the site and at every change in direction of the property boundaries. They usually bear on their tops the distance and orientation of property corners or boundaries. A typical survey beacon mark is S85°E, which shows that the property corner in question is located in the south but extends towards the east of the site by 85°. Survey beacons define the boundary and area of the site.

Elevations

These are the different heights on the surface of the site in relation to a standard reference point known as the bench mark (BM). The elevations of points within a site are always expressed in units above or below sea level. Locations which have equal elevations are joined together using contour lines. These lines help to define the topography of the land within a site.

Site orientation

This refers to a system of defining the site in terms of its direction to the north, south, east and west. Orientation is important in planning the building area to take into consideration factors such as the direction of rain, wind and sun within the site.

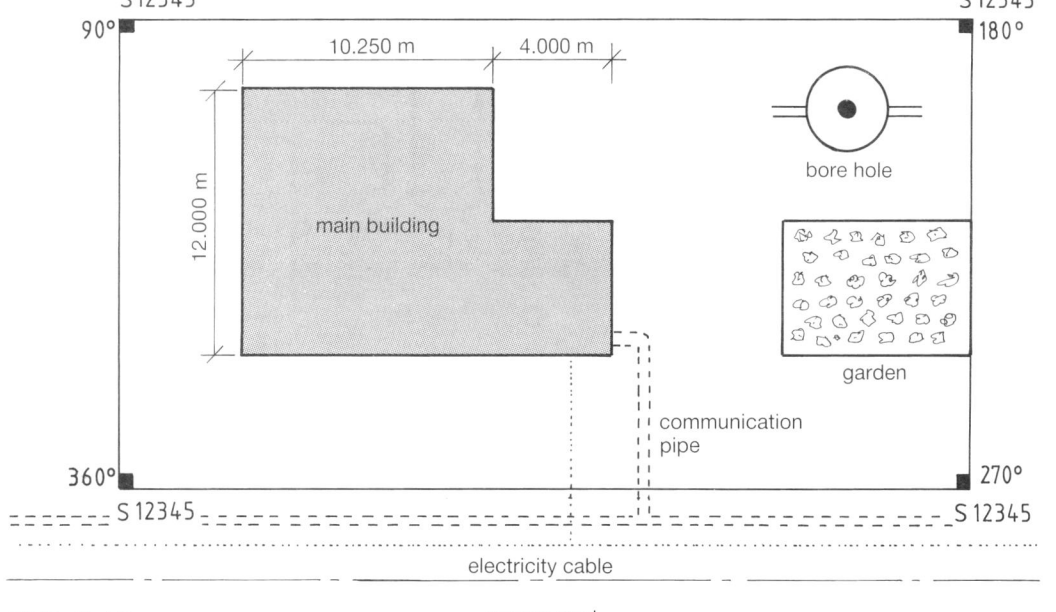

SITE PLAN

TECHNICAL DRAWING 3: BUILDING DRAWING

Physical features
These are permanent objects or features existing within the site or adjoining sites which are used for referencing or identification of the site. Such physical features may include existing buildings, trees, hills, roads, fences, churches etc.

Access road
A site chosen for building purposes requires an access road if there is none already or an indication of such a road if it exists. Therefore the site plan shows the means of reaching the site, whether such means is an expressway, a dual carriageway or a footpath.

Utilities
A site plan shows utility supply lines such as for water, electricity and gas. These features are shown on the site plan to indicate how they would be supplied to buildings within the site.

Procedure for drawing a site plan
1. Choose an appropriate scale that will contain the site drawing on available drawing paper. For small sites a scale of 1:200 is appropriate. However for larger sites scales of 1:500, 1:1000 or 1:2500 may be used.
2. Fasten your paper on to a drawing board and draw the border lines.
3. Draw the first property line in the correct bearing and scale.
4. Draw the next property line, working in a clockwise direction. Be sure that this next property line is in the correct bearing and scale.
5. Proceed as in step 4 until all the

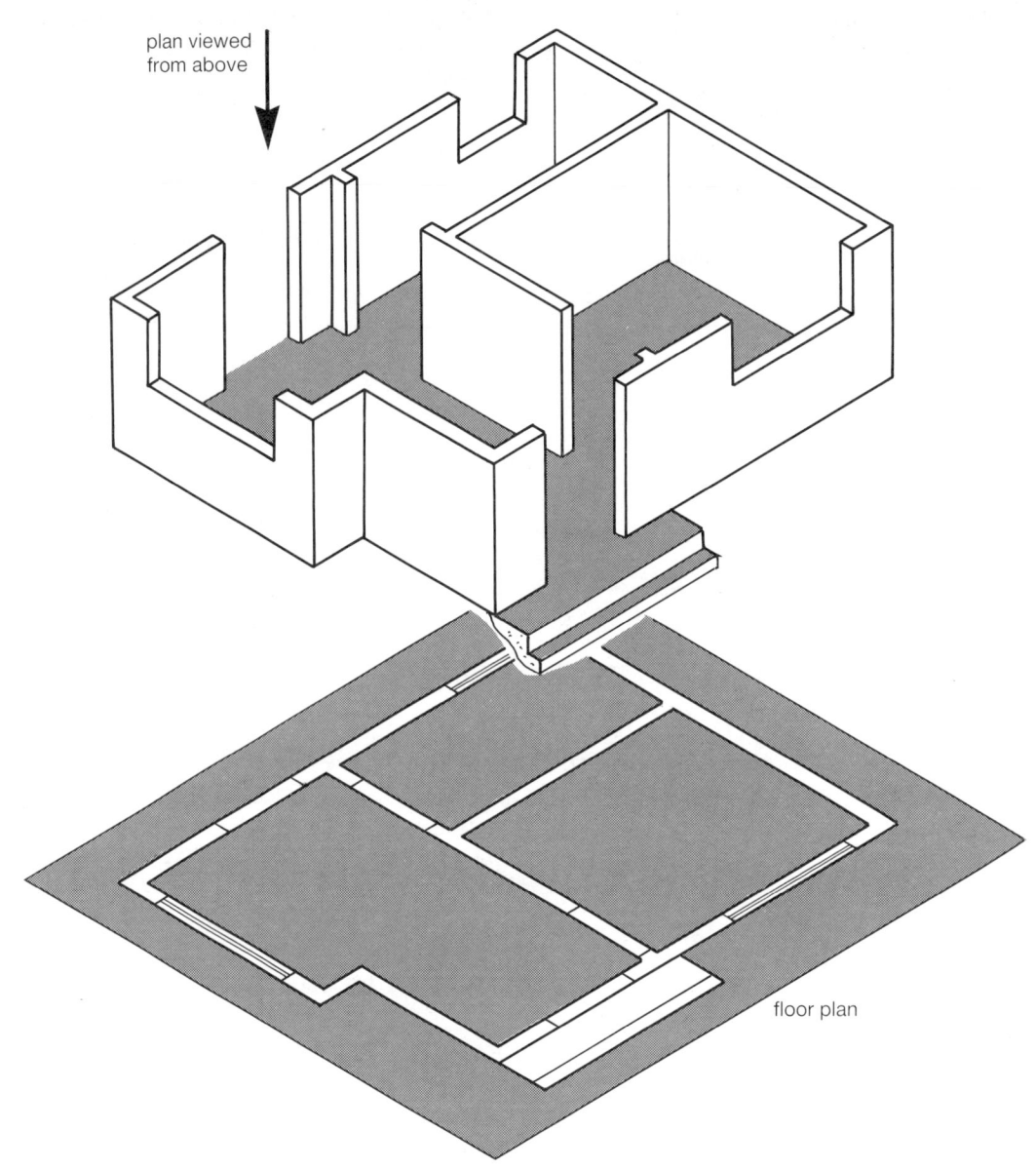

plan viewed from above

floor plan

property lines have been drawn and closed up to define the shape of the site.
6. Locate the important physical features on the site, using standard symbols. Where standard symbols are not used, legends must be provided to explain those used.
7. Draw lines to connect equal elevations on the site using information obtained from the site survey.
8. Determine the portions of the site to be developed or built upon, taking into account factors such as the topography of the land, vegetation, permanent physical features on the site and orientation.
9. Determine the location of the proposed building in relation to the direction of the sun, wind and access roads.
10. Draw the building line such that it reflects city or town planning regulations governing the area in which the site is located. It is recommended that the building line should not be less than 5 m from the centre of a footpath, 10 m from the centre of a dual carriageway and 15 m in the case of an expressway.
11. Allow at least 3 m between the building and other buildings or property boundaries. Also, do not use more than 60 per cent of the property area for your proposed building.
12. Indicate important features around the building. These include driveways, footpaths, septic tanks etc.
13. Indicate the orientation of the building or site. This helps to identify the location of the building.
14. Draw the title box and include the following information: scale, name of property owner, registration number of property, use for the proposed building and other required information.
15. Check your drawing to ensure that no useful information has been omitted.

Floor plan

Essentially, the floor plan of a building is a horizontal section through the window openings and doorways (see page 24). Thus, the floor plan shows the positions of windows, doors and all walls. In fact, the floor plan of a building can be regarded as the most important working drawing in the construction of a building. It enables the workman to know how many rooms a building should contain, the sizes of the rooms and the overall dimensions of the building. In multistorey building drawing, each floor has its own plan and these are designated as ground floor plan, first floor plan, second floor plan etc. (see right). Sometimes, however, two or more floor plans are identical. This usually occurs with floors above the ground floor. Where two plans are identical it is unnecessary to draw separate plans for each floor. The common practice is to state as a footnote: 'Third floor plan same as second floor' or 'second, third and fourth floor plans identical.'

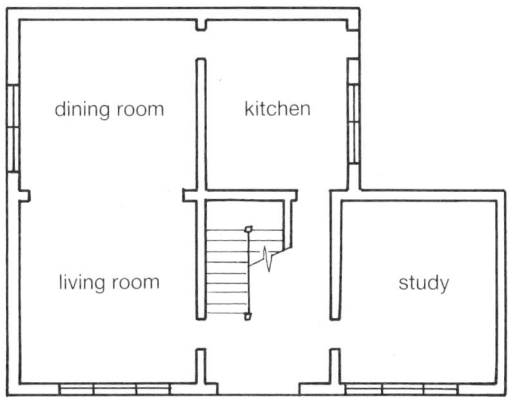

a) Ground floor plan

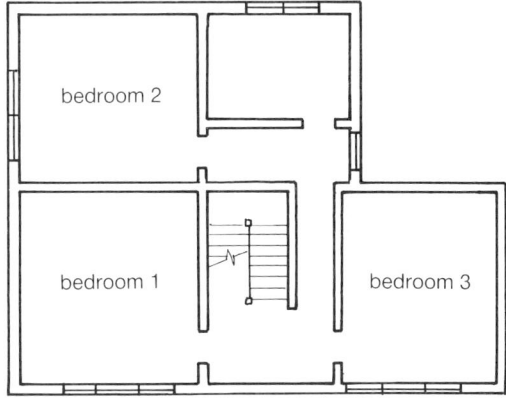

b) First floor plan

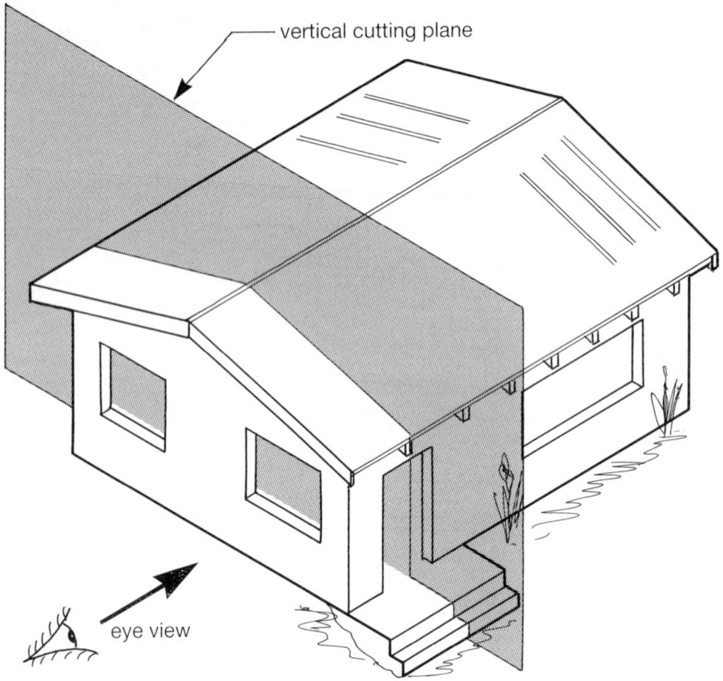

TECHNICAL DRAWING 3: BUILDING DRAWING

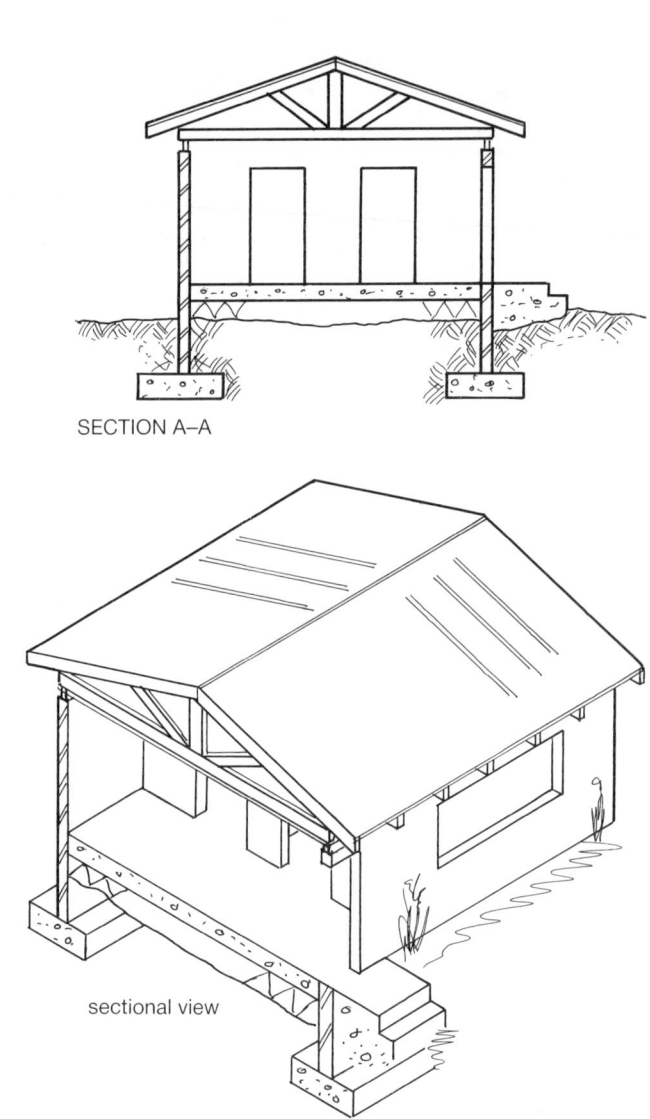

vertical cutting plane

eye view

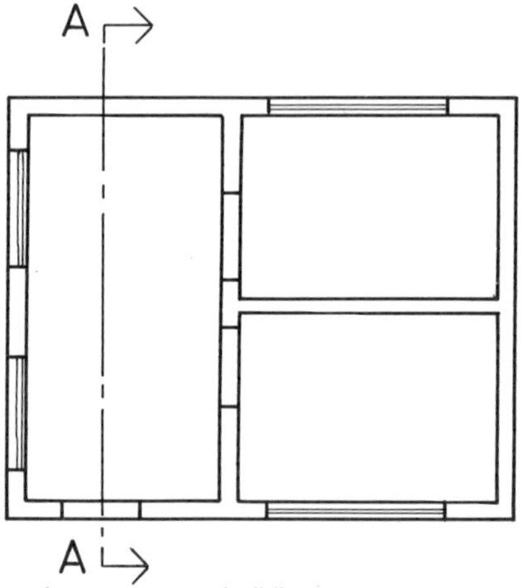

SECTION A–A

PLAN

sectional view

Offset section through a two-storey building

BUILDING CONSTRUCTION DRAWINGS

Sectional views of buildings

Sections of a building, obtained with the aid of vertical cutting planes, serve to show the construction of certain elements of the building: for example, floor heights, elevations (levels) of the floors, landings, windows, and so on.

Imagine a house cut by a vertical plane, and one side is removed. The remaining side is then projected to the profile planes of projection to reflect the sectional view of the house (see page 26). Cutting planes are indicated on the plan of a building by interrupted cutting plane lines with arrowheads at their ends which indicate the direction of the sectional view. Sections are usually designated by letters, for example: section A-A, or section Y-Y. A pictorial view of section A-A is shown.

Offset sections

Sometimes all the details of a building which should be revealed in a sectional view do not fall within the line of one cutting plane. In such situations a section is obtained by cutting the building with more than one plane. Such a section is called an offset section. In the drawing on the right for example, it is necessary to show the details of the main entrance door, the staircase and the toilet. An offset section AA is, therefore, taken to show the details needed.

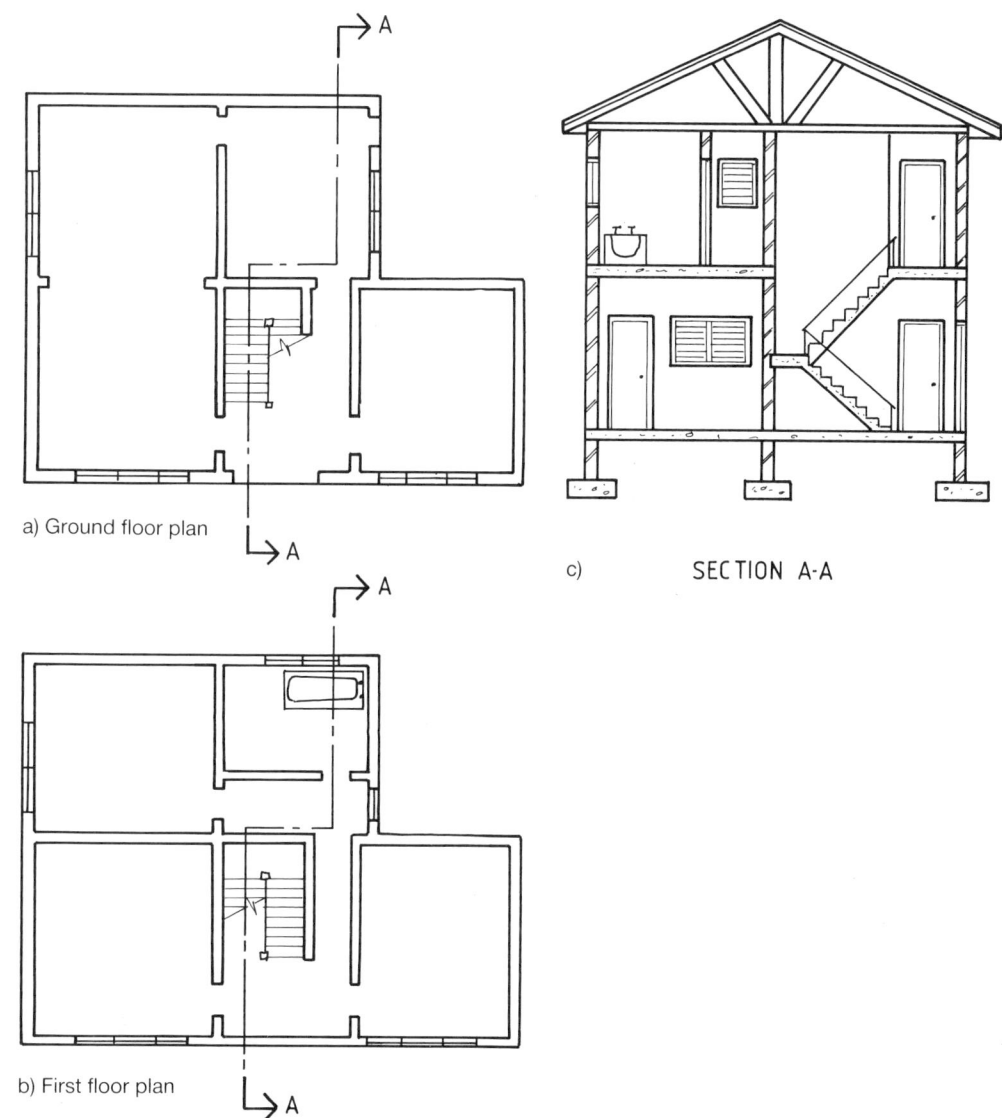

a) Ground floor plan

b) First floor plan

c) SECTION A-A

Room planning and specifications

Living room
The living room, in many homes, is the centre of activities. It is in the living room that guests are received and entertained. The living room often contains such household effects as the television, furniture and musical equipment.

A typical living room measures about 4.000 m × 5.500 m. Its location should be such that it will not be used as a passage to other parts of the house. The flat with the living room located as shown in the drawing on page 29 is regarded as planned well because it is separate from the other rooms.

Dining room
The dining room provides a special place for eating. A dining room measuring 3 m × 3 m can conveniently accommodate four people around a dining table. It therefore follows that the size of a dining room will be influenced by the number of prospective users. It is desirable to locate the dining room close or adjacent to the kitchen.

Bedrooms
Bedrooms are so important in a house that adequate thought should be given to their location, size and the number a house should contain. The major determining factor regarding the number of bedrooms in a building is the size of the family to occupy the building.

The three-bedroom flat or bungalow is very popular in many urban towns and is fairly adequate for most medium-sized families, that is, of four to five members.

Most authorities recommend that the minimum size of a bedroom should be 3 m × 3 m or 9 m^2.

As regards location, grouping bedrooms together in a wing or level of the house affords quietness and privacy. Bedrooms should be located close to bathrooms. Some bedrooms may have their own private baths. Each bedroom should have at least one entry door which opens into the hallway or passage.

Bathrooms
Every residential building requires at least one bathroom with a water closet. The bath or shower may be located in the same room as the water closet or in a separate room. Where the bath and water closet are located in the same room, they should occupy a space of not less than 1.500 m × 2.400 m. If in separate rooms, the shower or bath should occupy a space of not less than 1.200 m × 2.400 m and the water closet 0.900 m × 2.400 m. A wash basin should always be installed near the water closet.

Wardrobes or clothes closets
It is necessary for every bedroom in a house to have a wardrobe or closet. The wardrobe should be of a minimum size of 1.200 m span and 0.750 m deep. It is preferable to locate closets in internal walls rather than in external walls. Also, in order to conserve space, it is desirable to fix sliding doors to the wardrobes.

Kitchen
The kitchen provides space for cooking food and sometimes for eating. It should be located close to the dining room and should be adequately ventilated. The kitchen should occupy a minimum floor space of 3 m × 3 m.

Garage
A garage (not shown) to accommodate only one car should measure at least 3.500 m × 6.000 m, whereas one to accommodate two cars should be at least 7.000 m × 6.000 m. A door leading from the garage to the interior of the building may be useful.

One reason for preferring these sizes is that they involve little or no waste of ceiling boards, which usually come in 1.200 m × 1.200 m size.

BUILDING CONSTRUCTION DRAWINGS

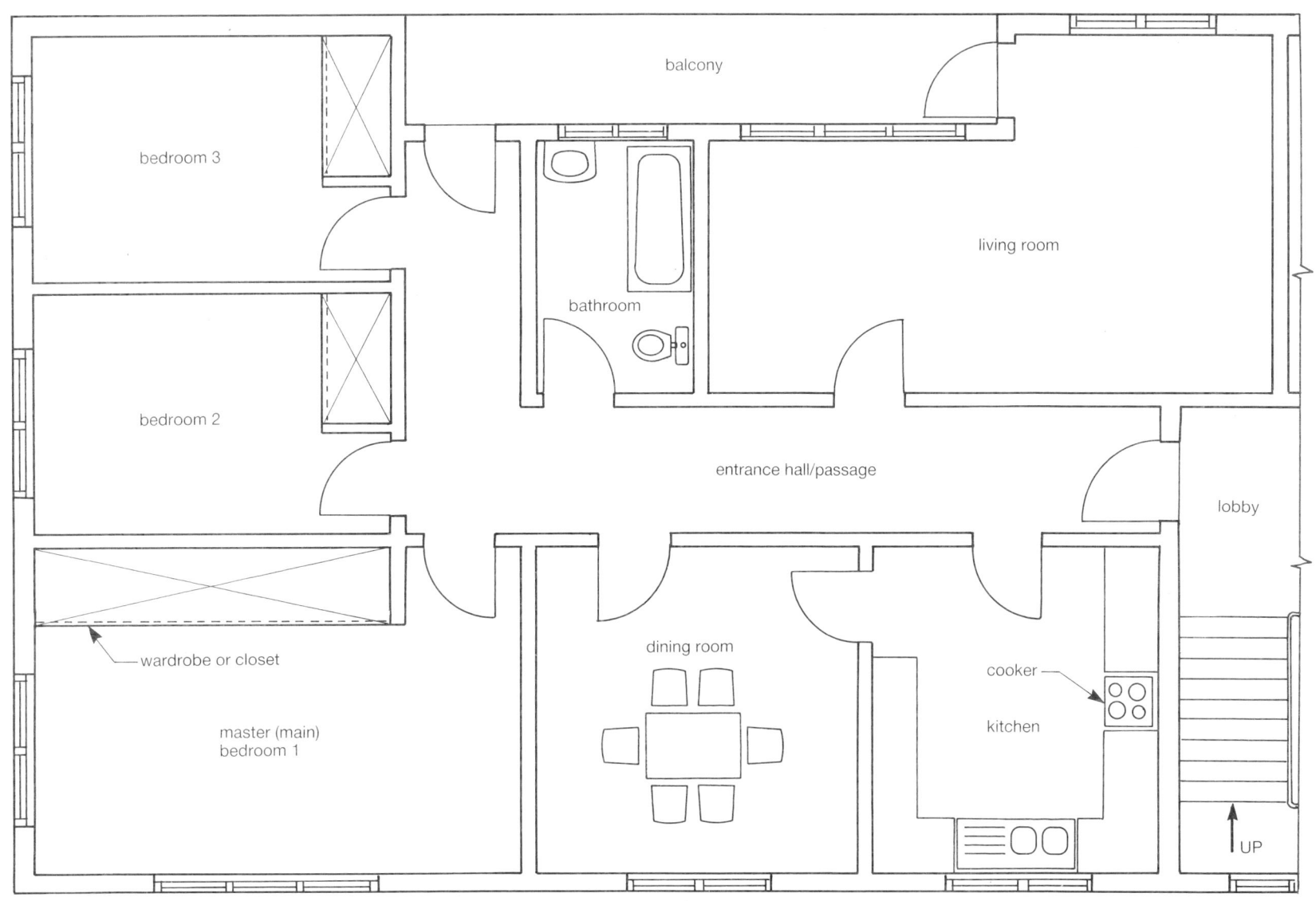

Plan of a three bedroom, first floor flat

29

TECHNICAL DRAWING 3: BUILDING DRAWING

Layout of rooms

When planning a building it is important to avoid certain rooms becoming passageways. The top building is badly planned because the living room is the only route to all the other rooms in the bungalow. In the bottom bungalow the living quarters are separated from the sleeping area and the use of one room as a general passageway is reduced.

Exercises

1. Design a two-bedroom house that has the following: one toilet, one bathroom, a living room, a kitchen and a garage. Be sure to use the correct sizes of each of these spaces in your design.
2. Refer to the badly planned bungalow on the right. Examine the layout and correct it so that is looks better planned. Do not copy the example below it.
3. Draw a section X-X of the well planned bungalow on the right.

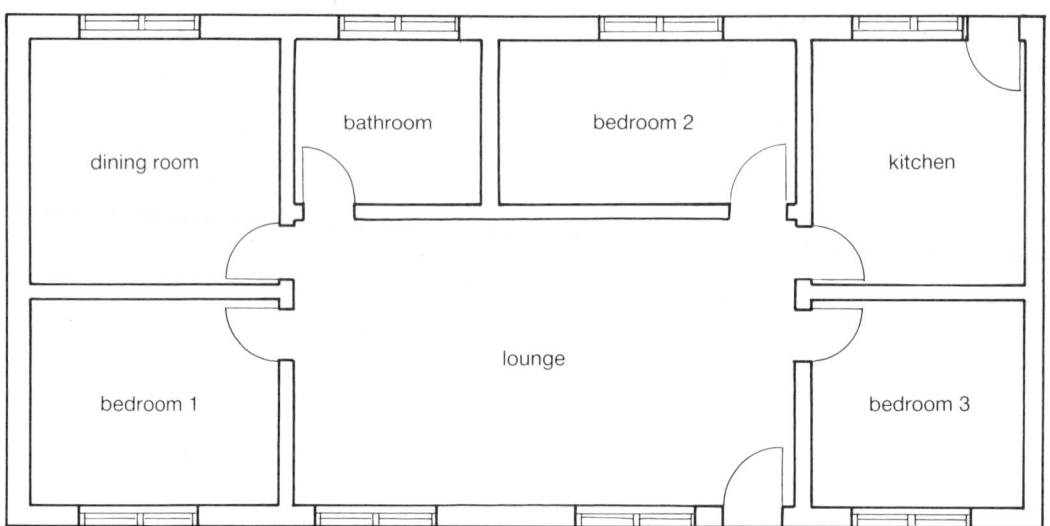

A badly planned bungalow

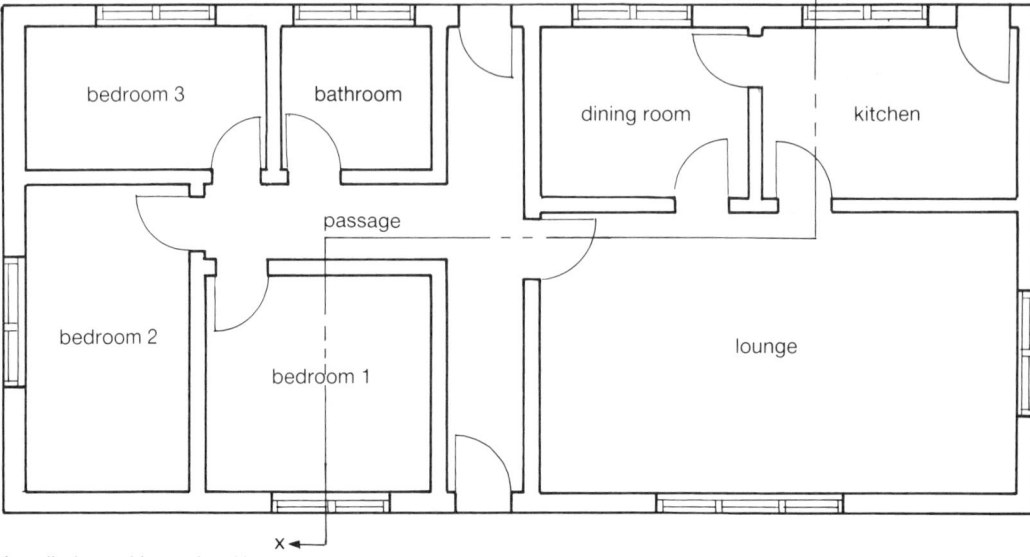

A well planned bungalow X

5. Foundations

Different types of foundation are designed for different buildings depending on the nature of the soil on which the building is to be erected, the type of building and the nature of the load the soil is expected to carry. Whereas a soft clayey soil, for example, would require a very deep strip foundation, a rocky soil would not require a foundation as deep (see below).

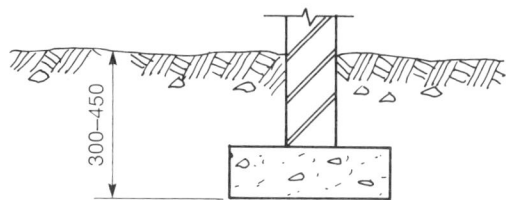

Shallow foundations in rock

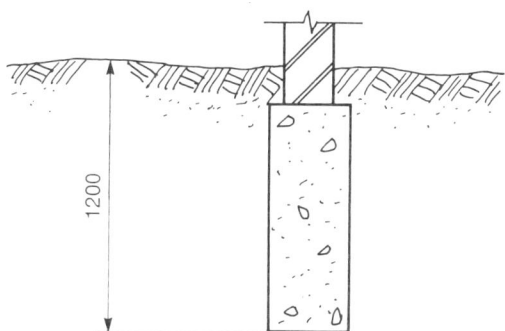

Deep foundations in soft clay

Similarly, whereas a factory in which heavy duty operations would be carried out will require a raft foundation (see bottom drawing on page 33), a bungalow on an identical type of soil may need only a simple strip foundation.

Depth of concrete in foundations

The major factor which determines the depth to which concrete should be placed in a foundation is the nature or type of soil. For a rocky soil the depth could be as shallow as 300 mm; for a soil of soft clay or in marshy areas the depth could be up to 1200 mm or more if a strip foundation is to be used.

Strip foundation

The strip foundation (see below) is the commonest type of foundation the student is likely to meet both in drawing and at construction sites. It is adopted on one- to two-storey (and sometimes up to four-storey) buildings on firm non-shrinkable subsoils, such as gravel or laterite. A strip foundation usually consists of a continuous strip of concrete which provides a continuous ground bearing for load-bearing walls.

Determining graphically the size of the concrete base

It has been found that fractures in concrete strip foundations usually occur in a direction corresponding to an angle of 45° from the face of the wall. If this fracture occurs in the foundation, providing the dimension P is equal to D (see drawing on page 32), the area of the concrete base which bears on the soil is not reduced and the soil below it can still safely carry the load.

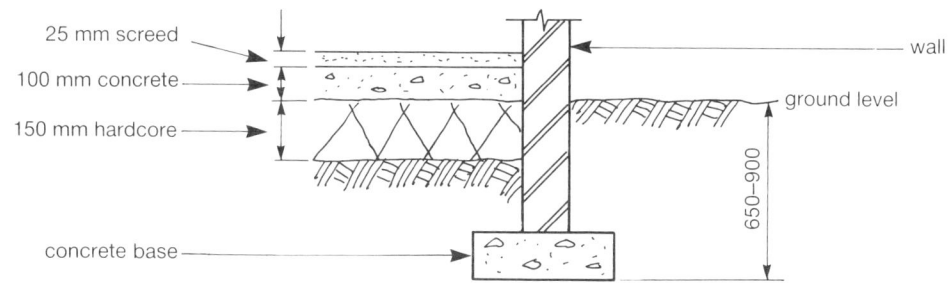

Strip foundation

TECHNICAL DRAWING 3: BUILDING DRAWING

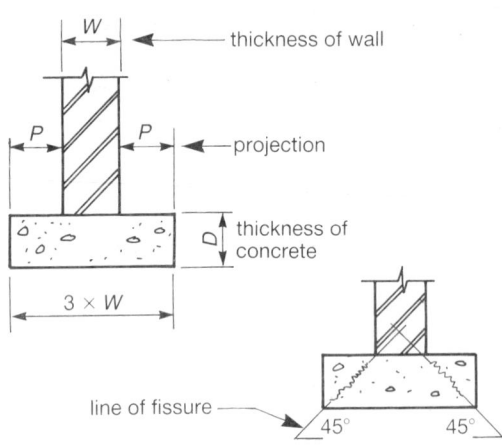

Size of concrete base

The above phenomenon has been found very useful in determining graphically the dimensions of the concrete base of a strip foundation.

Given the width of the wall as W, the strip foundation is drawn as follows:
1. Draw two parallel lines that define W as the width of the wall as shown.
2. Draw a base line to cut the two parallel lines in step 1. Project this base line on either side of the parallel lines.
3. Draw lines at 45° from O and C.
4. Mark distance OB equal to the thickness of the wall W. Let this distance be P. From the point B drop a perpendicular to cut the oblique (45°) line. The intersection of the perpendicular line and the oblique line at D determines the depth or thickness of the concrete in foundation.
5. The drawing is then completed by projecting the line through D to the left side of the wall, and dropping a perpendicular line from the point E to intersect the line through D at F. The rectangle BDFE with the width of wall OC centrally located, shows the depth of concrete in foundation BD (or EF) and the width of foundation BE (or DF).

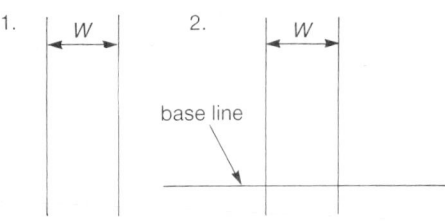

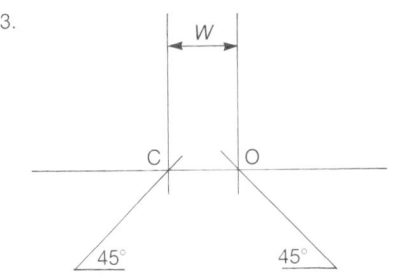

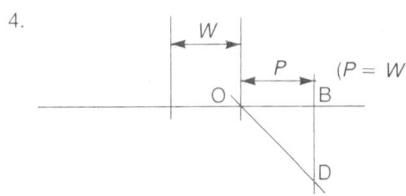

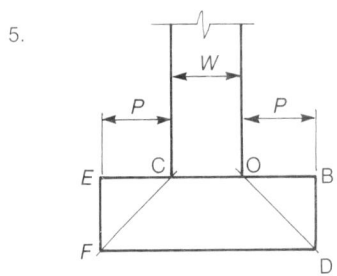

Drawing a strip foundation

Other types of foundation

The design and construction details of the strip foundation can be varied to reflect soil conditions. The deep strip foundation (see below) is adopted on clayey or marshy soils. The depth of the concrete base is increased while the width is reduced, thus reducing excavation costs.

The wide strip foundation, on the other hand, is employed when the soil has low bearing capacity such as in marshy and built-up soils. This type of foundation

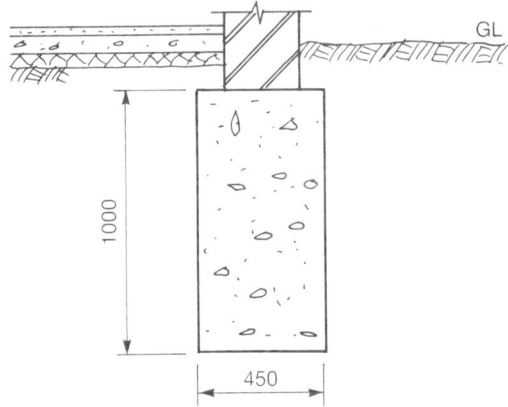

Deep strip foundation

32

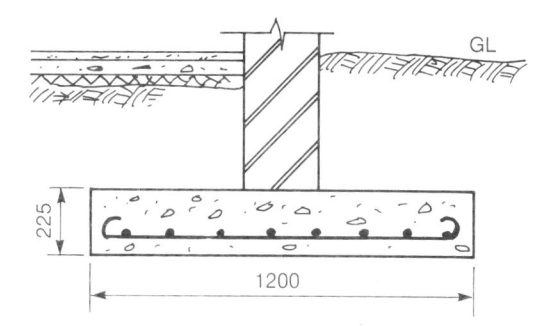

FOUNDATIONS

Wide strip foundation

enables the load to be spread over a wide area. The concrete base is usually reinforced to withstand tensile forces that may be set up at the bottom side of the foundation due to the large width.

Pad foundation
These are generally isolated foundations consisting of a block of reinforced concrete meant to support reinforced concrete or steel piers or columns, especially in framed structures. The thickness of the concrete pad is usually not less than 150 mm.

Raft foundation
A raft foundation covers the entire area of a building and is, therefore, not a separate unit from the floor slab. The raft may extend beyond the external walls of the building. Usually a raft foundation consists of suitably reinforced concrete and may be up to 300 mm thick.

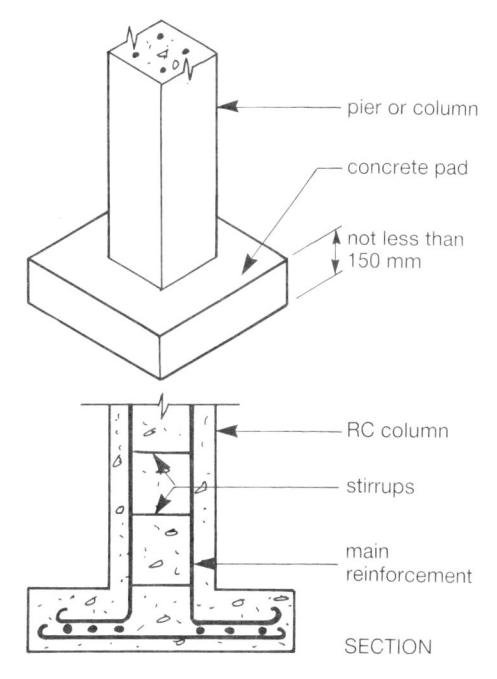

Pad foundation

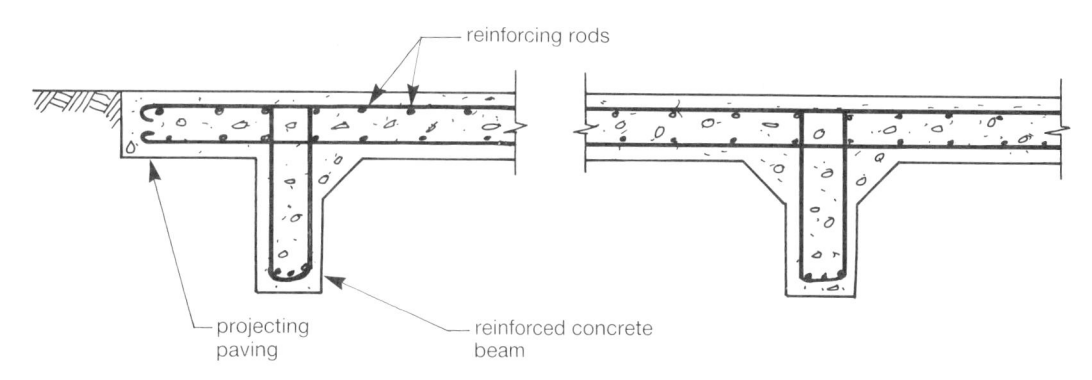

Raft foundation

Exercises

1. Draw a foundation plan of the bungalow whose plan is shown on page 30.
2. Using graphical methods, determine the depth of concrete in foundation for the walls and width of foundations indicated below:
   a) 225 mm, 900 mm
   b) 225 mm, 750 mm
   c) 225 mm, 675 mm
   d) 150 mm, 900 mm
   e) 150 mm, 750 mm
   f) 150 mm, 450 mm
   Use a scale of 1:10.
3. Draw a section Y-Y of a strip foundation with the following data:
   a) Depth of foundation = 900 mm
   b) Depth of concrete in foundation = 180 mm
   c) Width of foundation = 750 mm
   d) Width of wall = 150 mm
   Use appropriate symbols and conventions to represent each material.

33

6. Floors

Floors are broadly classified into ground floors and upper floors.

Ground floors

Ground floors may be solid or suspended. Solid ground floors are more economical to construct and consist of concrete laid on solid ground. The two methods used in constructing solid ground floors are monolithic construction and separate construction.

In the monolithic method the screed is laid within three hours of placing an *in situ* oversite concrete, that is, before the concrete sets. In separate construction the screed is applied after the concrete base has set and hardened. The construction details of a solid ground floor are shown below.

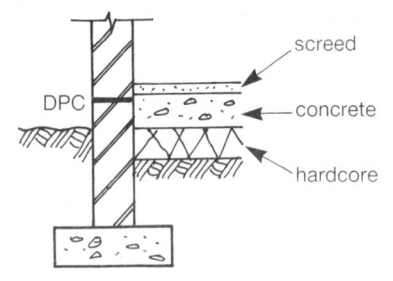

a) Concrete blockwork

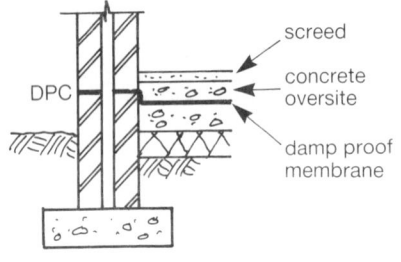

b) Cavity wall brickwork

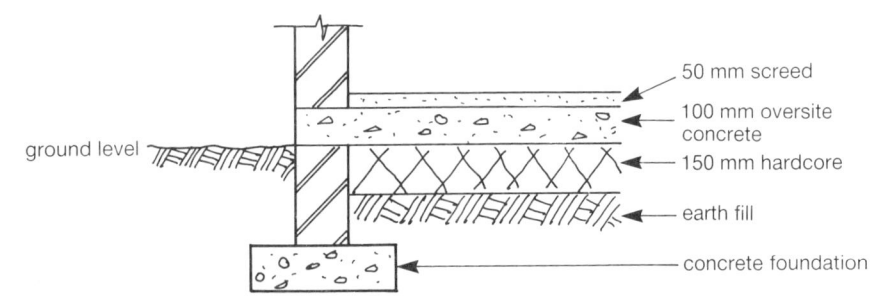

c) Section through a solid ground floor

Alternative methods of constructing solid ground floors are shown on the left. In c) the use of oversite concrete laid into the wall makes the use of an additional damp proof course (DPC) unnecessary. In a) damp proof courses have been provided to keep off moisture from the ground. The oversite concrete in b) was laid in two layers with a damp proof membrane (DPM), e.g. mastic asphalt, bitumen or plastic sheets, between the layers.

Steps in drawing a solid ground floor

The following procedure should be a guide when drawing solid ground floors:

1. Decide on the scale to use. Note that in an examination situation a scale may be given and this must be strictly followed. Ordinarily, a scale of 1:50 or 1:20 may be used.
2. Draw a horizontal line defining the ground level (GL) at a convenient position on the drawing paper (see a), page 35).
3. Draw two perpendicular parallel lines across the GL line, the distance between the parallel lines being the thickness of the wall (see a), page 35). Let the intersections of the perpendicular lines and the GL line be O and P, respectively.

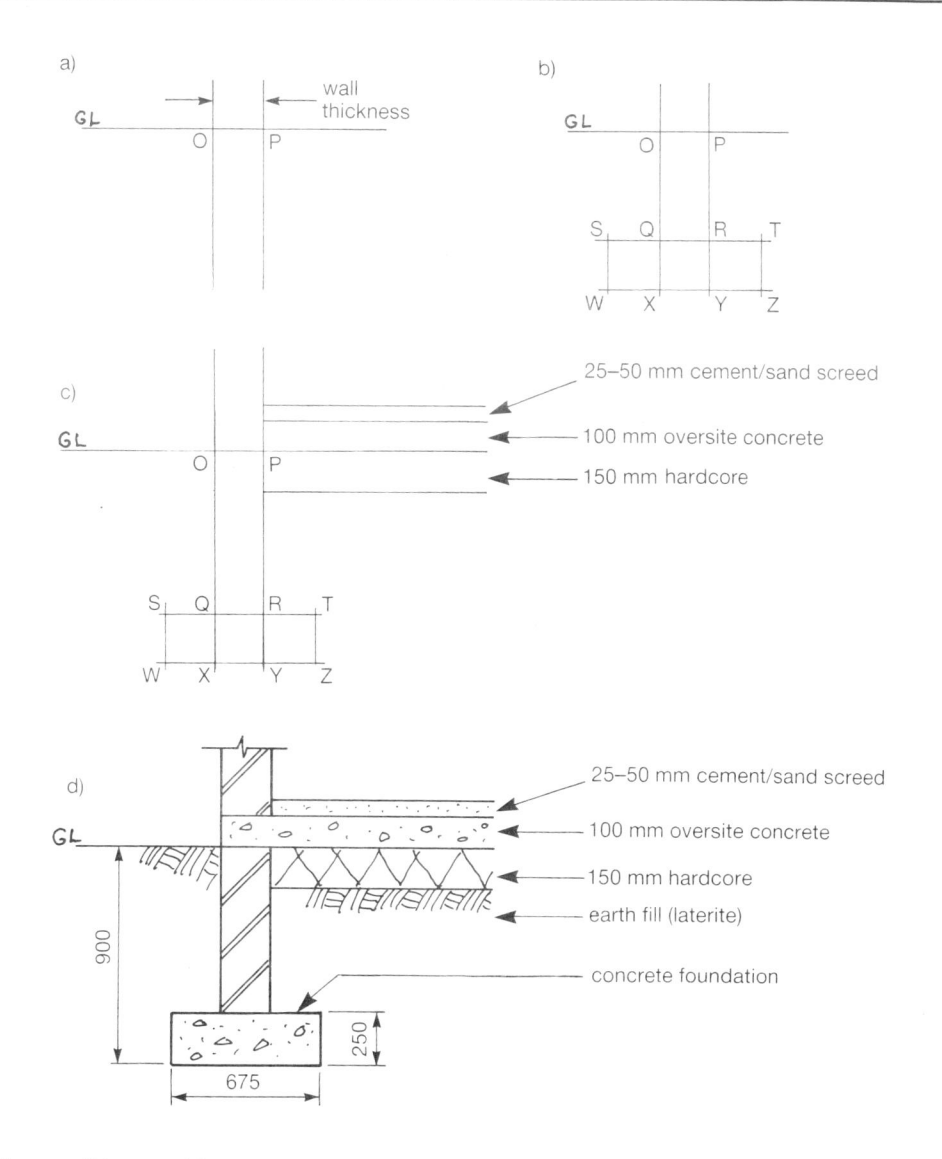

Drawing a solid ground floor

4. From either of the points, O or P, measure downwards along the parallel lines to get the depth of the foundation (from the ground level). Let this distance be OX or PY (b).
5. From X and Y measure out (upwards) the thickness of the concrete in foundation, XQ and YR. Also, from Q and R mark out QS and RT equal to the projection of the concrete base on either side of the wall. The concrete foundation is thus represented by SWZT (see (b)).
6. Mark out the floor elements as follows: 150 mm below the ground level for the hardcore; 150 mm from the lower end of the hardcore line as the earth fill; 100 mm above the ground level for the oversite concrete and 25–50 mm above the concrete floor for the screed. These are shown in c). Note that if a separate DPC is to be used, it should be placed about 225 mm above the ground level.
7. Erase all construction lines.
8. Using appropriate symbols hatch the wall and represent concrete, hardcore, earth fill and screed as shown in d).
9. Dimension the drawing.

Suspended ground floors

A suspended ground floor is usually constructed of timber. It consists of timber boarding nailed to floor joists that are supported on the wall plates. The drawing on page 36 shows the construction details of a suspended ground floor. The procedure for drawing a suspended ground floor is similar to that adopted for a solid ground floor. The major difference is that there is

TECHNICAL DRAWING 3: BUILDING DRAWING

A suspended ground floor

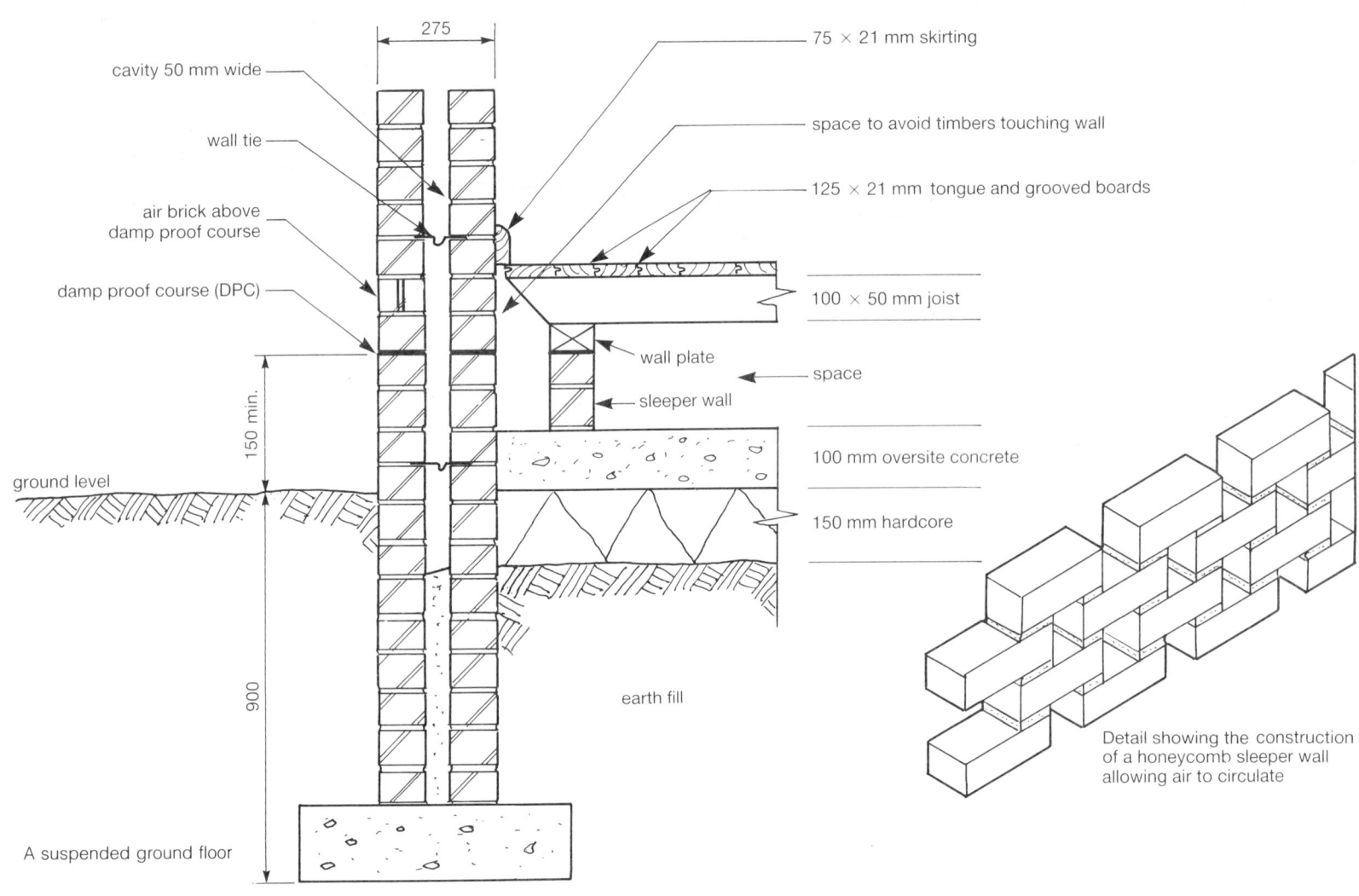

Labels on drawing:
- 275
- cavity 50 mm wide
- wall tie
- air brick above damp proof course
- damp proof course (DPC)
- 150 min.
- ground level
- 900
- 75 × 21 mm skirting
- space to avoid timbers touching wall
- 125 × 21 mm tongue and grooved boards
- 100 × 50 mm joist
- wall plate
- space
- sleeper wall
- 100 mm oversite concrete
- 150 mm hardcore
- earth fill

Detail showing the construction of a honeycomb sleeper wall allowing air to circulate

no screed on the oversite concrete of a suspended floor; rather, timber members are built on to the floor. The dimensions and construction details of these members are shown.

A suspended floor should be constructed on a ground whose surface is covered with a layer of concrete not less than 100 mm thick and supported on a bed of hardcore. The space between the upper surface of the concrete and the underside of the wall plate should not be

FLOORS

less than 75 mm. The space between any other timber suspended on the wall plate (e.g. the floor joist) and the upper surface of the concrete should not be more than 125 mm.

Sleeper walls should be built on the surface concrete, the bricks being laid in a honeycomb pattern to permit adequate ventilation. An illustration of a honeycomb construction is shown on page 36.

The air bricks placed in the external walls ensure a current of air through the space beneath the floor. The floor joists are usually 100 mm x 50 mm, spaced at about 400 mm centres and nailed to the wall plate. Boards laid over the joists are usually tongued and grooved (see right) to prevent draughts and dirt blowing into the room.

Each board is usually 100 mm or 125 mm wide and 25 mm thick. The edges of the floorboards should be 13 mm from the surrounding walls to prevent damp penetration. The gap is closed using a skirting (see left).

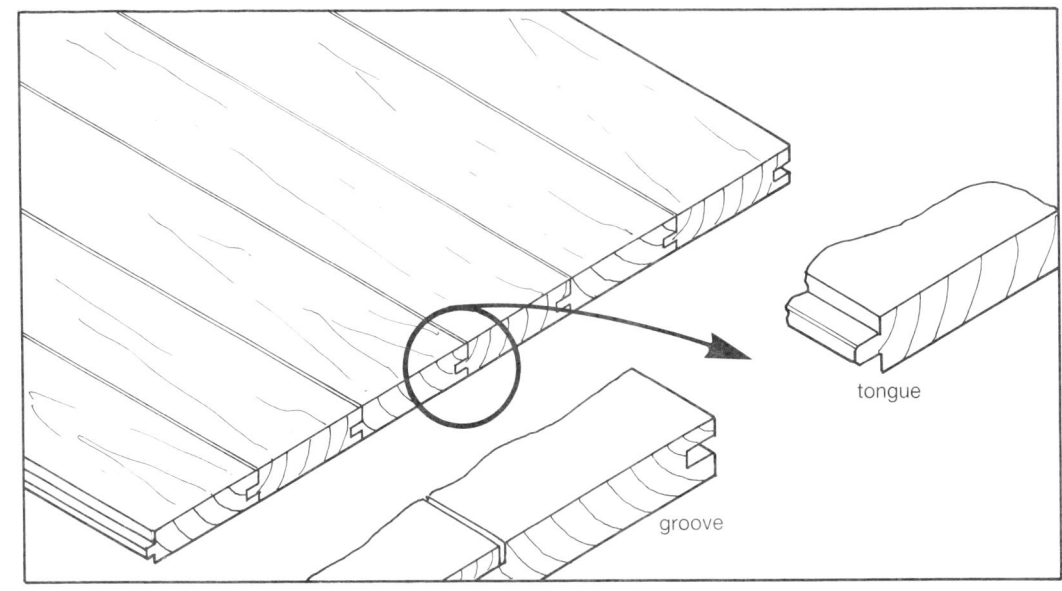

Tongue and grooved joints for floorboards

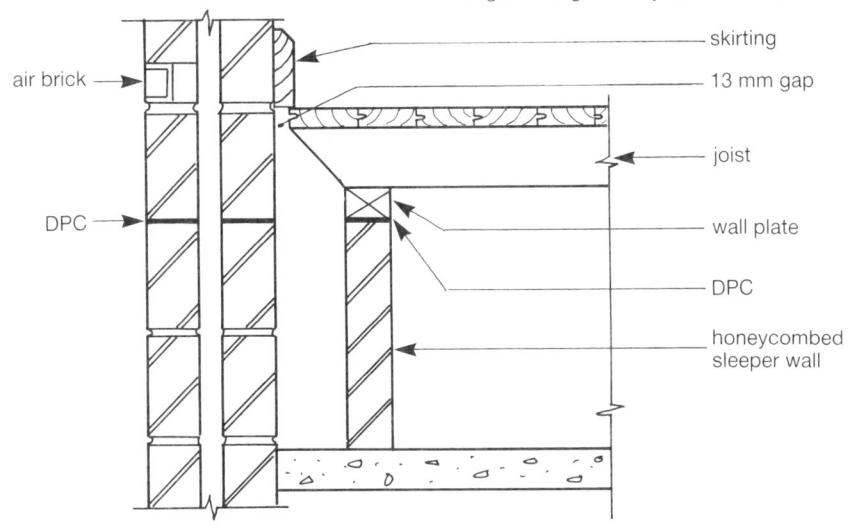

Construction details of a suspended floor

Upper floors

Although timber is still used for the construction of upper floors of some special buildings, it has been largely replaced by concrete. The method of construction of timber upper floors is similar to that of suspended ground floors, except that timber upper floors have joists which are usually longer and deeper. Some limitations of timber upper floors include poor fire-resistance, low load-bearing capacity and poor sound absorption.

37

TECHNICAL DRAWING 3: BUILDING DRAWING

Reinforced concrete floors

In constructing a reinforced concrete (RC) floor (see page 39), the concrete is laid in one solid mass, to a thickness of between 100 mm and 300 mm. The concrete is reinforced both in the transverse and the longitudinal directions. Timber formwork and centering (temporary timber boarding laid horizontally) are constructed to assist in supporting the wet concrete.

It is common practice to make the centering span the external walls, with allowance provided for continuing the vertical reinforcements for columns from

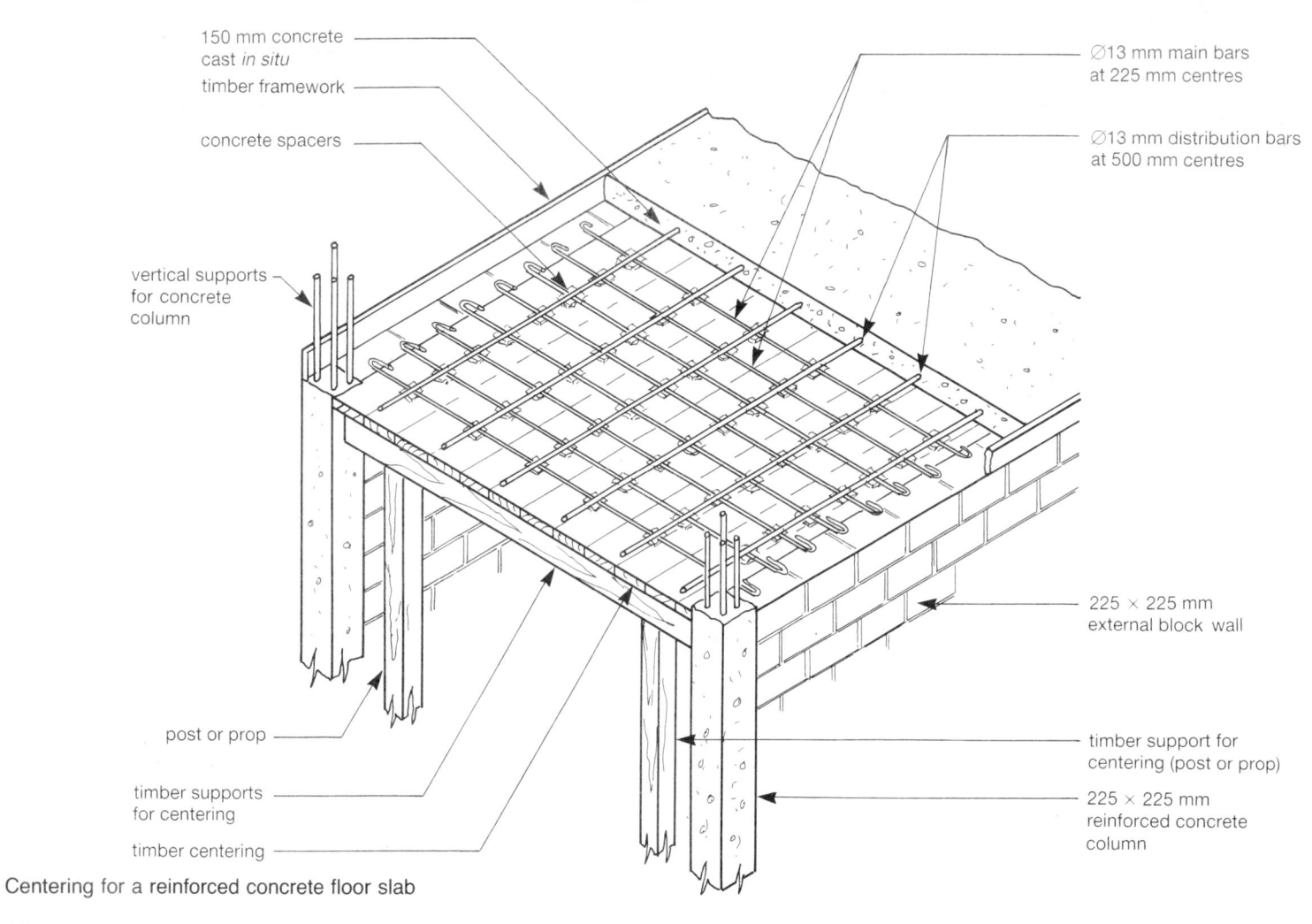

Labels:
- 150 mm concrete cast *in situ*
- timber framework
- concrete spacers
- vertical supports for concrete column
- post or prop
- timber supports for centering
- timber centering
- ⌀13 mm main bars at 225 mm centres
- ⌀13 mm distribution bars at 500 mm centres
- 225 × 225 mm external block wall
- timber support for centering (post or prop)
- 225 × 225 mm reinforced concrete column

Centering for a reinforced concrete floor slab

38

FLOORS

the ground floor (see page 38). The centering is supported in between the walls by timber props or posts.

The main reinforcements are laid on top of the centering by means of small concrete cubes or blocks called spacers (or biscuit bones, locally) which are tied to the reinforcing bars. These spacers are used in providing concrete cover to the steel rods in order to prevent rusting. The main reinforcement usually consists of 13 mm-diameter mild-steel rods spaced from 150 mm to 225 mm centres, and these span the floor between the external walls. The size of the distribution rods across the main reinforcement ranges from 6 mm to 13 mm in diameter and are spaced at between 450 mm and 900 mm centres. The distribution rods are tied to the main bars using binding wire.

Tensile forces in floors of large spans can be reduced by constructing beams across the walls. Details of a beam cast with a floor slab are shown.

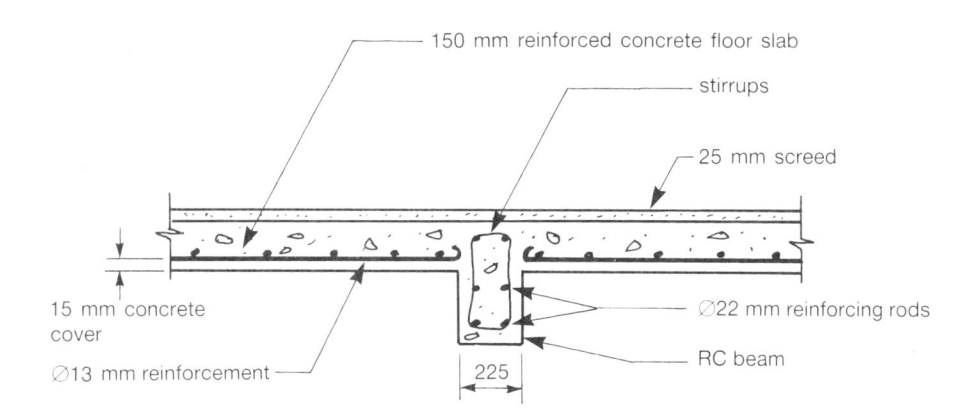

Reinforced concrete upper floor

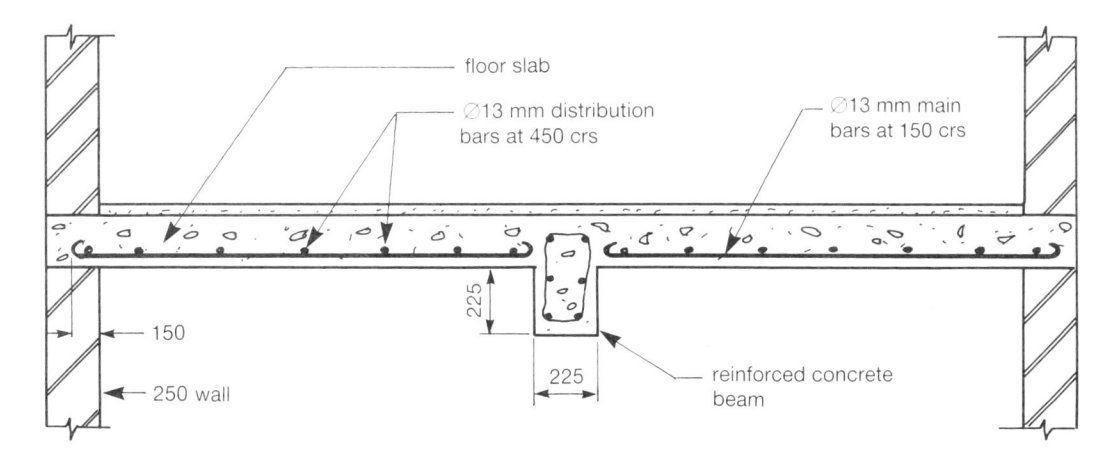

Beam construction with a floor slab

Exercises

1. Draw a section through a solid ground floor, showing all materials and their appropriate conventions.
2. With the aid of neat sketches, show the location of the DPC and DPM:
   a) in a concrete block wall,
   b) in a cavity wall.
3. Draw a section through a reinforced concrete floor to reflect the following details:
   a) the location and spacing of main reinforcement rods,
   b) the location and spacing of distribution rods,
   c) the construction detail of the floor into the walls,
   d) the depth of the concrete cover,
   e) other members of the RC floor.
4. Show by means of appropriate sketches the construction details of a reinforced concrete upper floor.

7. Wall openings

In general, an opening is created in a wall so that a door or a window can be fitted into it. The drawing below left shows a typical wall opening and its features.

Jambs and reveals

The term jamb as used in brickwork or blockwork refers to the full height of an opening on either side of the window. Reveal is the term used to describe the thickness of the wall revealed after cutting the opening in the wall. The reveal shows the surface of the brickwork all through the height of the opening, as shown.

Jambs are sometimes recessed or rebated (see below right) to accommodate door or window frames with large cross-sections, in order to reduce the amount of frame visible to the eye. Another reason for rebating jambs of openings is to ensure a more rigidly fixed door or window frame.

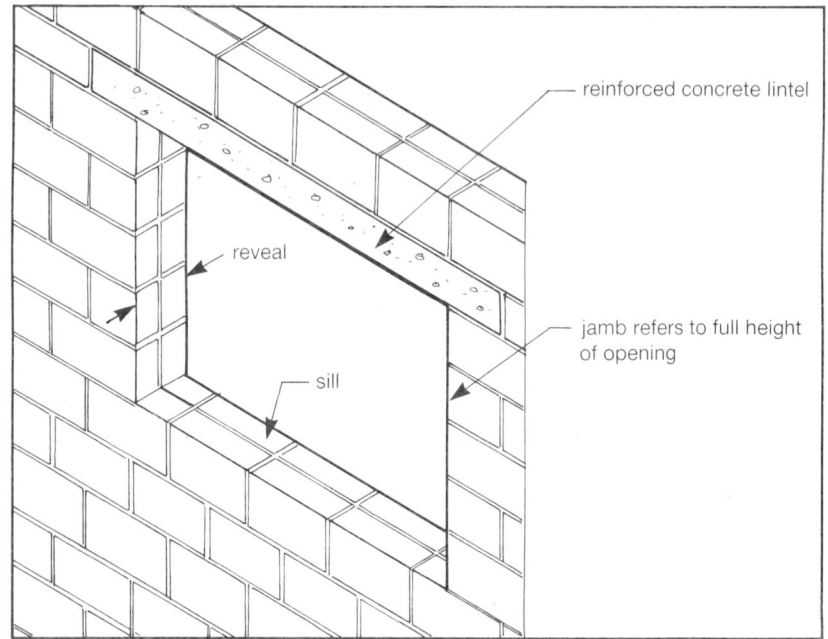

A typical wall opening

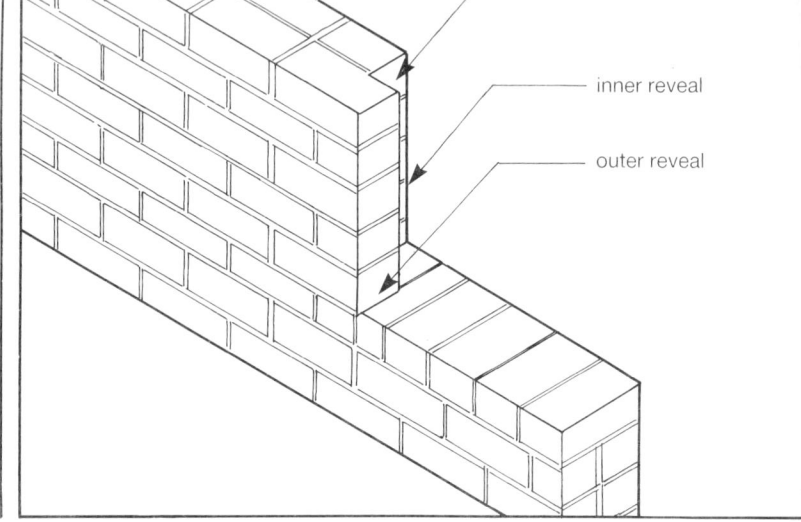

Rebates in jambs

WALL OPENINGS

Window sills

A sill is the member at the base of a window opening. The purpose of a sill is to throw off water and to form a covering for the wall at the base of the opening. In addition, a sill gives the opening a finished appearance. Materials from which sills are made include stone, concrete, brick, tile and timber. In order to prevent water entering from the sill to the wall, an anticapillary groove or throating and, sometimes, a steel water bar are provided under the sill. In addition a sill should have sufficient slope to throw water off the wall.

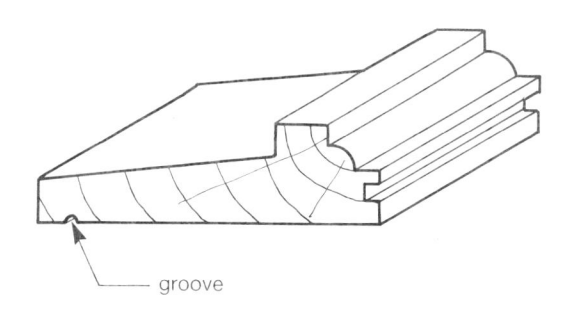

Section through a window opening

- putty
- 75 × 100 mm hardwood sill
- throating
- concrete sub-sill
- damp proof membrane
- 6 × 25 mm steel water bar
- glass
- 44 × 63 mm casement bottom rail
- 25 mm window board
- inner leaf of cavity wall
- cavity

A hardwood sill

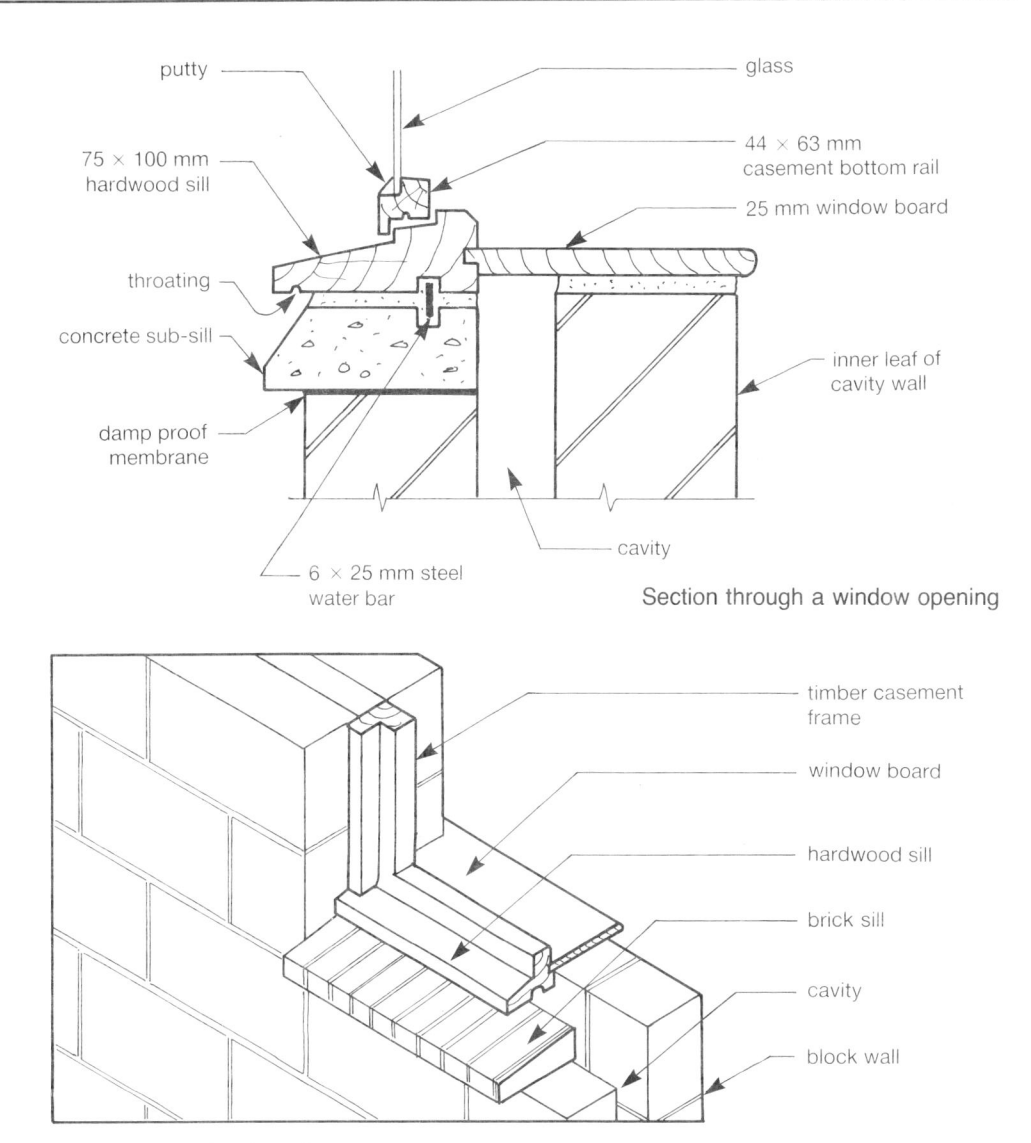

- groove

Section showing details of a sill

- timber casement frame
- window board
- hardwood sill
- brick sill
- cavity
- block wall

TECHNICAL DRAWING 3: BUILDING DRAWING

Lintels

Door and window openings are sometimes bridged with a single unit that spans the opening. This device is known as a lintel (sometimes spelt lintol). A lintel supports the wall over it.

A lintel can be made of wood, brick or concrete. Most lintels are now made of reinforced concrete as shown here. The ends of the lintel must be built into the brickwork or blockwork over the jambs so as to transmit the weight to the walls. The portion of the brickwork or blockwork on which the ends of a lintel rest is termed the bearing.

One desirable quality which concrete lacks when used for lintels is tensile strength. This is because concrete is strong in compression but weak in tension, and hence it must be reinforced. The reinforcement rods can be placed in the upper part but *must* be placed in the lower part of the lintel to resist tensile forces.

The steel rods are placed at a minimum of 25 mm up from the bottom of the lintel to give protection against corrosion. Pictures c) and d) also show alternative methods of arranging the reinforcement rods in a lintel. For spans larger than 1200 mm the lintel design in c) is used, but for spans less than 1200 mm it is more economical to adopt the method in d).

The ends of the reinforcement rods are usually hooked (picture b)) in order to provide a better bond between the concrete and the reinforcement, as well as to provide greater strength to the lintel.

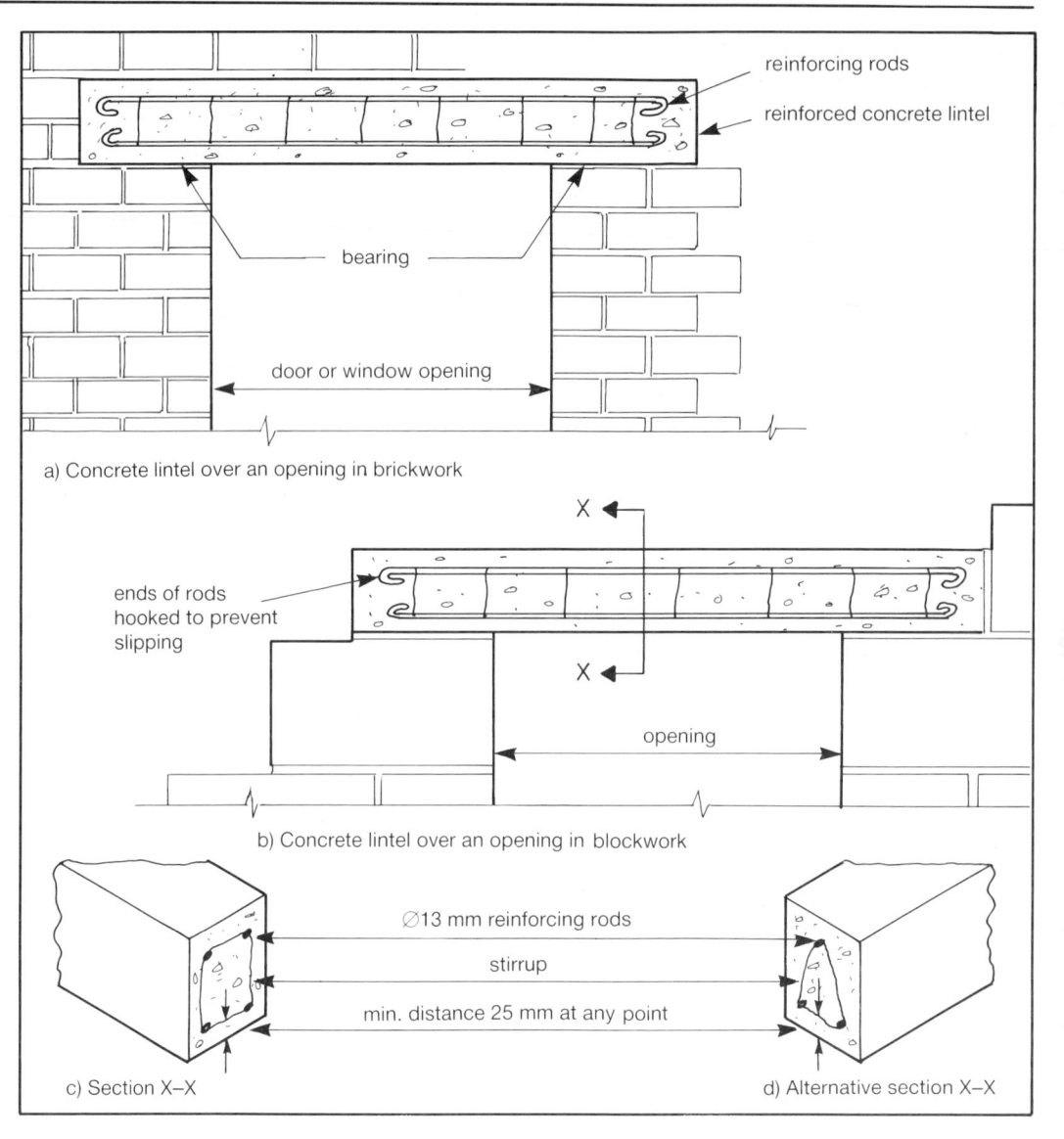

a) Concrete lintel over an opening in brickwork

b) Concrete lintel over an opening in blockwork

c) Section X–X

d) Alternative section X–X

Reinforcements in a lintel

Arches

Often, the openings in walls of public buildings are bridged with arches. For example, arches are often used to bridge openings of gates and main entrances to large buildings, town halls, religious buildings and colleges. The photograph shows the archways into the Dome of the Rock mosque in Jerusalem.

Traditionally, arches are features of stonework or brickwork, but arches are today often made in concrete on site or precast.

There are three classes of brick arches: gauged, axed and rough. These derive their names from the finished shape of the bricks used in the construction of the arches.

TECHNICAL DRAWING 3: BUILDING DRAWING

Gauged arches
These are arches whose bricks are accurately prepared to a wedge shape and are usually employed when high-quality work is desired.

To draw a gauged arch
1. Draw the centre line.
2. Draw the springing line to intersect the centre line at right angles.
3. Draw the lines of the skew-backs.
4. Produce the lines of the skew-backs to intersect the centre line of the arch. The point of intersection locates the centre from which all the arch joints will radiate.
5. Measure out the depth of the arch from the springing line above the centre line. Let this depth be AB (see (c)). Through point B draw a line to intersect the skew-back lines at X_1 and X_2.
6. Divide the line $Y_1 Y_2$ into a number of convenient parts depending on the span of the arch and the thickness of each brick. Number these divisions.
7. Join the centre O to each of these divisions by producing each line to touch the line $X_1 X_2$.
8. Complete the exercise by drawing the walls around the arch as in a).

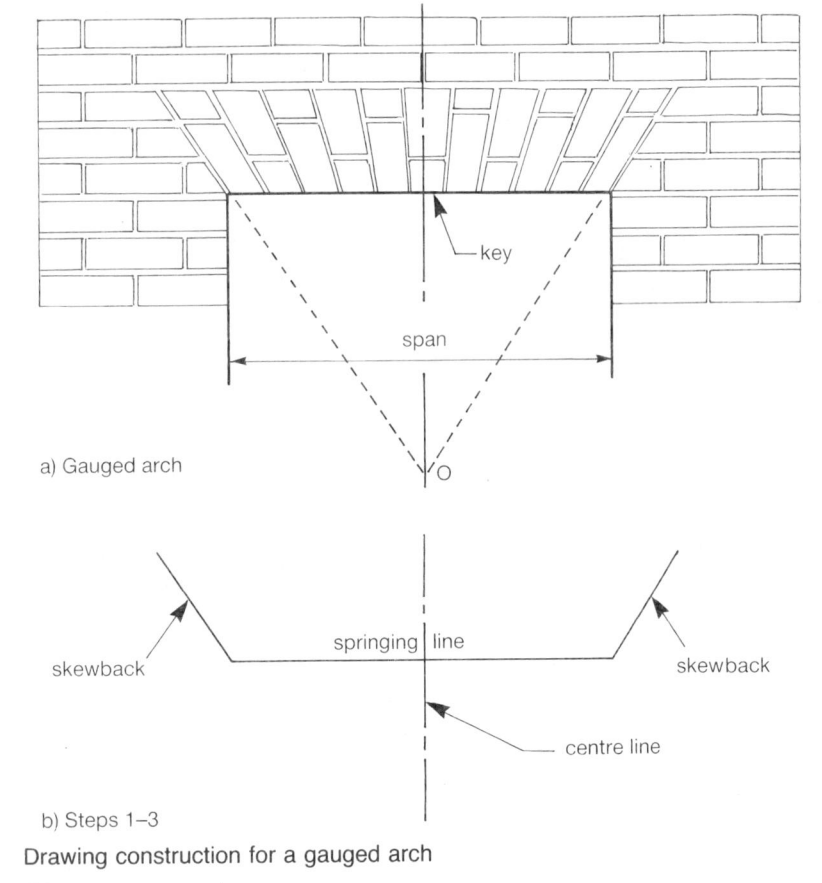

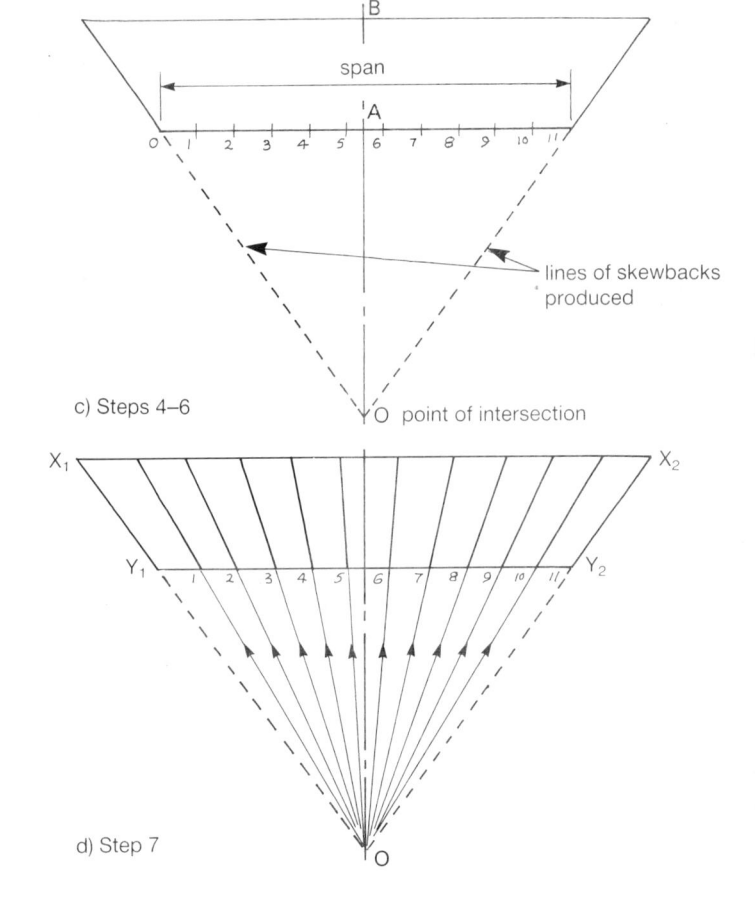

a) Gauged arch

b) Steps 1–3

c) Steps 4–6

d) Step 7

Drawing construction for a gauged arch

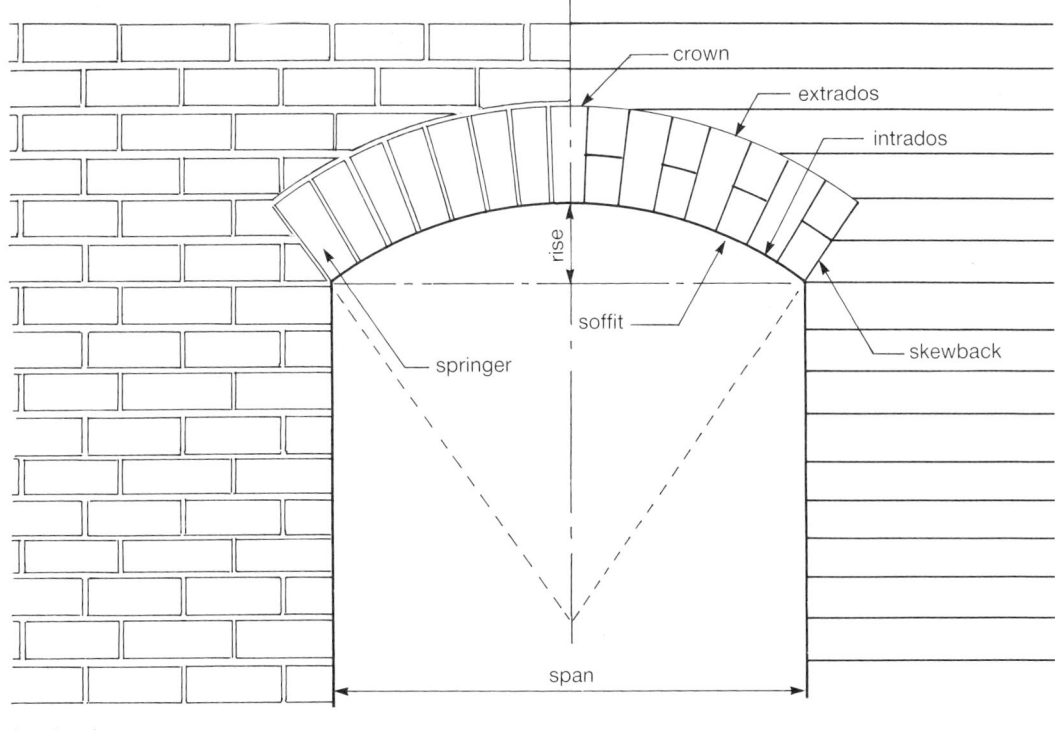

WALL OPENINGS

Axed arch

Drawing construction for an axed (segmented) arch

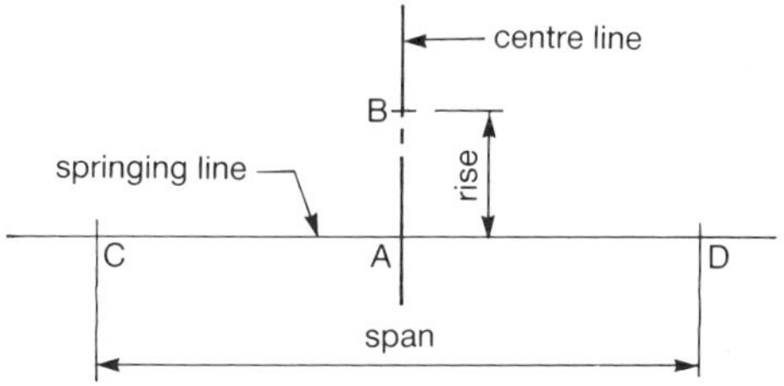

Axed arches
These can be segmental (see above) or semicircular (see page 46).

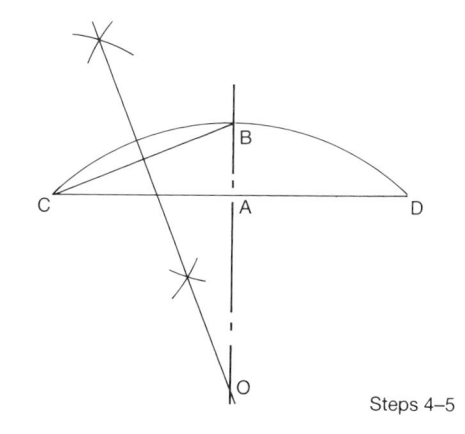

To draw an axed segmental arch
1. Draw a horizontal line to represent the springing line.
2. Draw the vertical centre line.
3. Set out the rise on the centre line. Let the line of the rise be AB.
4. Join the point C on the springing line to B. AC is half the span of the arch.
5. Bisect BC. The centre of the arch, curve O, lies at the point of intersection of the bisector and the centre line.
6. The curve CD represents the intrados of the arch. Also from centre O, draw a curve for the extrados to give the depth of the arch.
7. Complete the drawing by adding the bricks and walls as shown above.

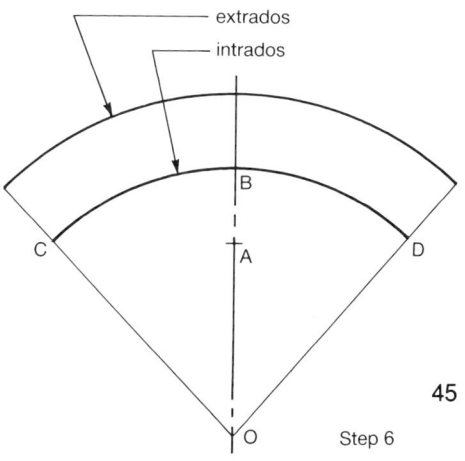

Steps 1–3

Steps 4–5

Step 6

45

TECHNICAL DRAWING 3: BUILDING DRAWING

To draw an axed semicircular arch

1. Draw the centre line.
2. Draw the springing line.
3. With the radius half the span, and the focal point at the intersection of the centre line and the springing line, draw the soffit semicircle. Note that the rise could also be used to form the radius of the semicircle.
4. Increase the radius of the semicircle in step 3 by the depth of the arch (one brick length in this case).
5. With the same focal point as in step 3 describe another semicircle to obtain a complete arch.
6. Draw the walls and bricks as shown.

Axed arches are constructed using good facing bricks which have been cut to a wedge shape with the aid of a template and a cutting tool known as a skutch.

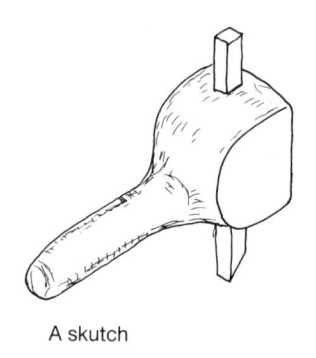

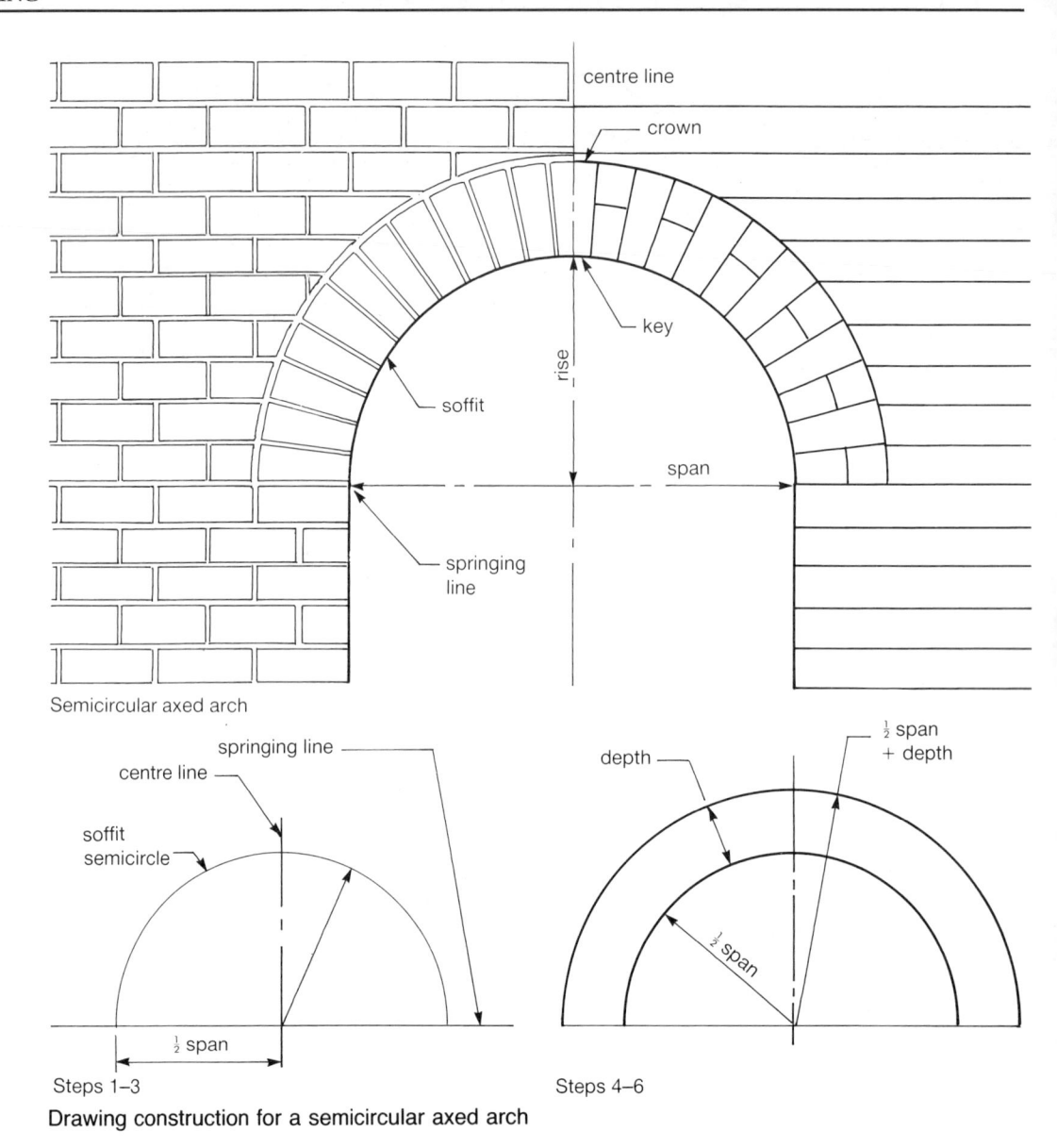

A skutch

Semicircular axed arch

Steps 1–3

Steps 4–6

Drawing construction for a semicircular axed arch

WALL OPENINGS

Rough arches
When an arch is built with common bricks without any cutting and with wedge-shaped joints between bricks it is said to be rough. Rough arches are set out from their bottom edges.

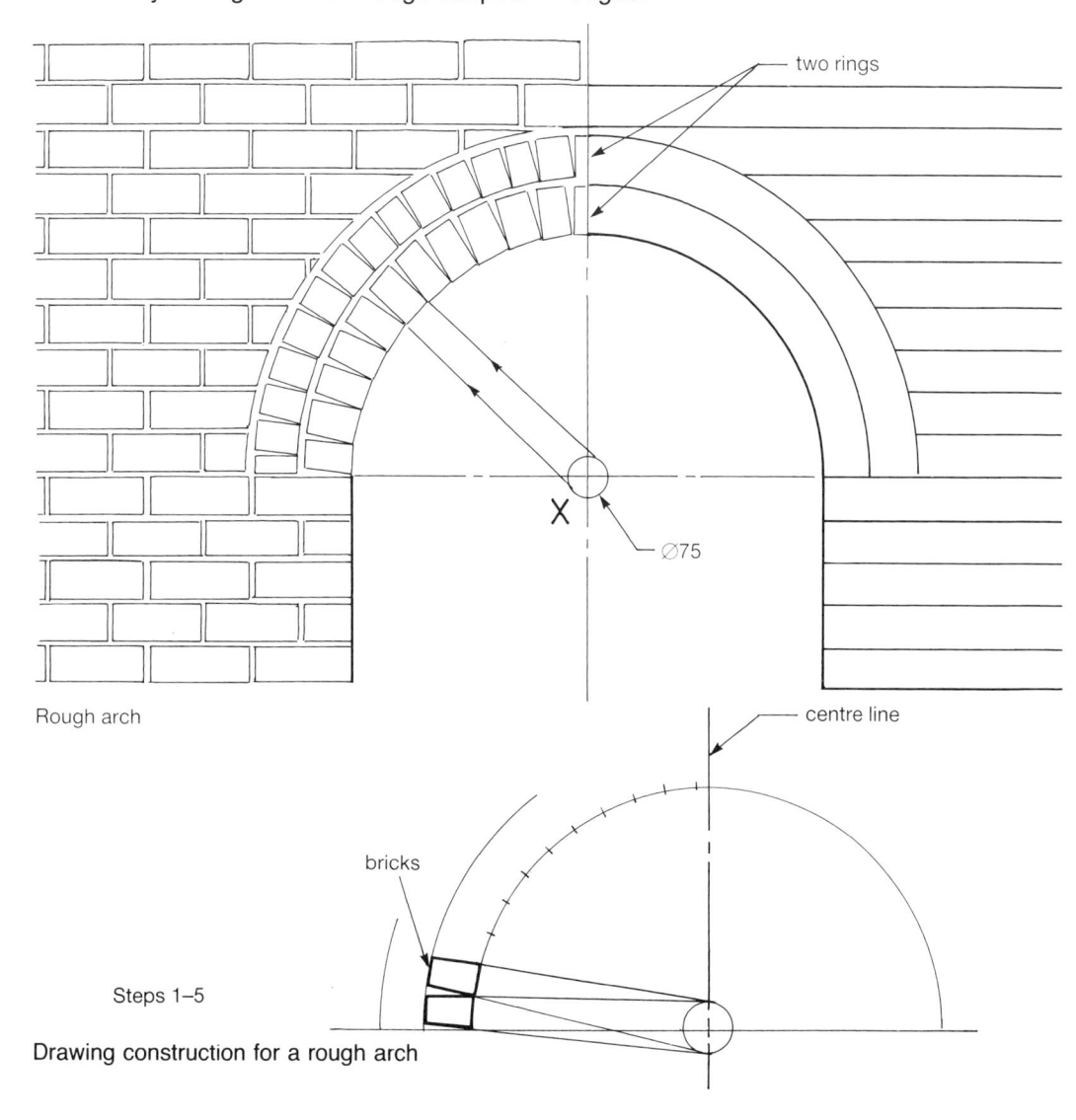

Rough arch

Steps 1–5

Drawing construction for a rough arch

To draw a rough arch
1. Draw the centre line and the springing line.
2. Describe a circle of diameter 75 mm using the intersection of the centre line and the springing line, X, as the focal point. Note that the actual size of this circle will depend on the scale being used for the drawing.
3. With a radius equal to the rise of the arch and centre, X, describe the semicircle of the soffit of the arch.
4. Set out distances equal to the diameter of the small circle (75 mm) to represent the widths of the bricks.
5. Join the points produced in step 4 tangentially to each side of the 75 mm circle, producing the line to define the parallel sides of each arch brick.
6. Repeat the process for the second ring of the arch.
7. Draw the walls around the arch.

Exercises
1. Identify and make a sketch of each of the following:
   a) reveal,
   b) jamb,
   c) sill,
   d) lintel.
2. Draw a section through a lintel, showing all appropriate materials and other construction details.
3. Draw a gauged arch; label its parts appropriately.

8. Doors and windows

Doors

A door is usually secured or hung to the reveal of an opening by means of metal hinges fixed to a solid wood or metal frame. A door together with its frame and hinges is referred to as a doorset.

Fixed door frames

Door frames may be of timber or steel. Timber frames may be secured:
1. by using galvanised steel brackets (see below left) or
2. by driving 150 mm nails into the back of the frame at course heights and building the nails into the wall (see below right).

Metal door frames are secured by lugs which are built into the wall. Some manufacturers include these lugs with the doorset.

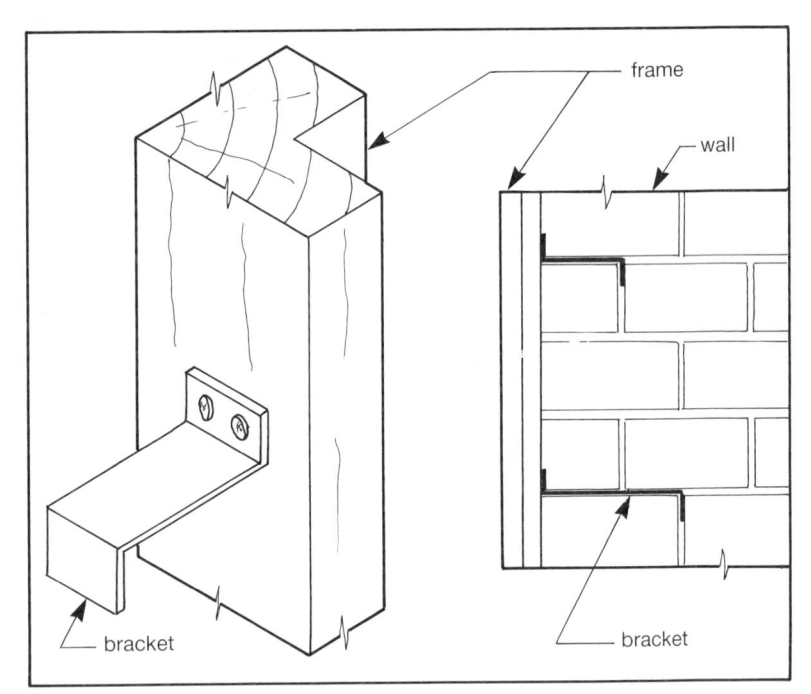

Galvanised steel bracket

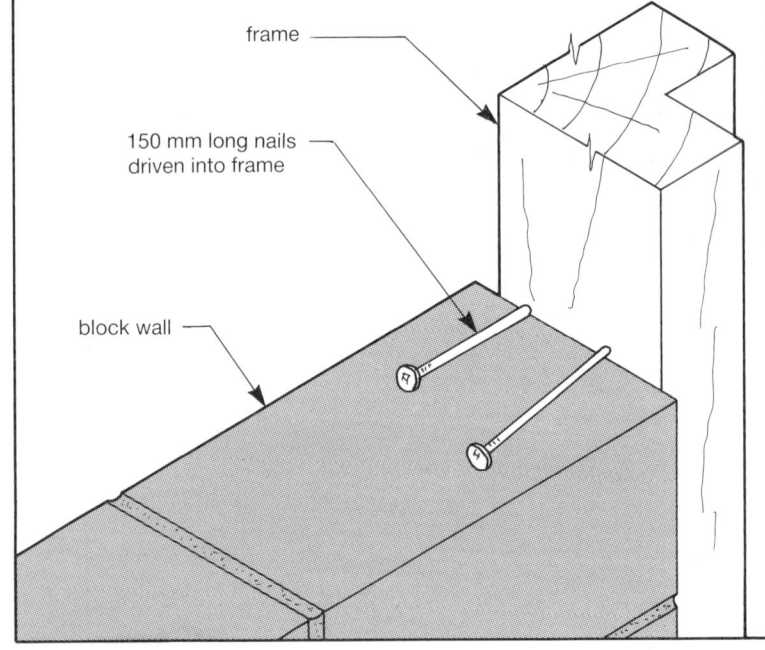

An alternative method using nails

DOORS AND WINDOWS

Door sizes

Doors come in a wide range of sizes. However, the standard sizes include:

526 mm × 2040 mm
626 mm × 2040 mm
726 mm × 2040 mm
826 mm × 2040 mm

Usually about 74 mm and 60 mm are added to the width and height, respectively, when determining the dimensions of the wall opening to accommodate each size of door.

Classification of doors

Doors may be classified as panelled doors, flush doors or matchboarded doors.

A six-panelled door

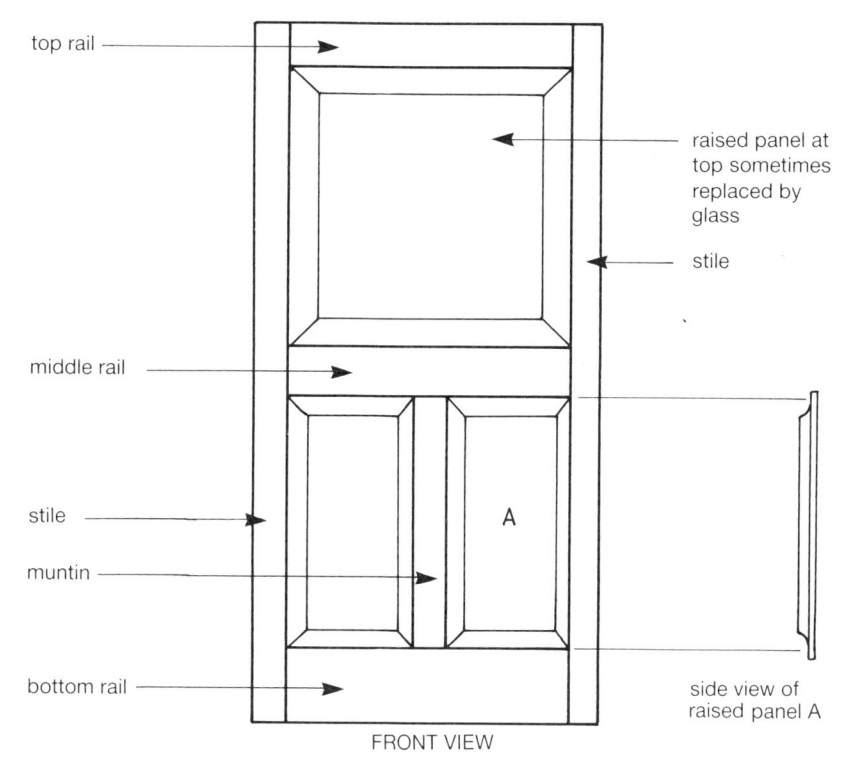

Features of a raised three-panelled door

Panelled doors

These can be described as one-panel, two-panel, three-panel etc., up to six-panel. Panelled doors (see photograph) consist of panels, stiles, top, bottom and middle rails, and sometimes muntins, which are vertical intermediate pieces tenoned to the top, bottom or middle rails.

Panels to doors can be of glass, plain plywood, plain solid timber or raised solid timber. Raised panels are more attractive but have the disadvantage of allowing dust to settle on them.

Apart from the panels, all other members of a panelled door are grooved at the edges to receive one or more panels. The panels may be lined with mouldings or left square. The members (stiles, rails and muntins) are joined together using mortice and tenon joints, dowelled joints, or pinned mortice and tenon joints (see page 50).

49

TECHNICAL DRAWING 3: BUILDING DRAWING

Dowel joints for door frames

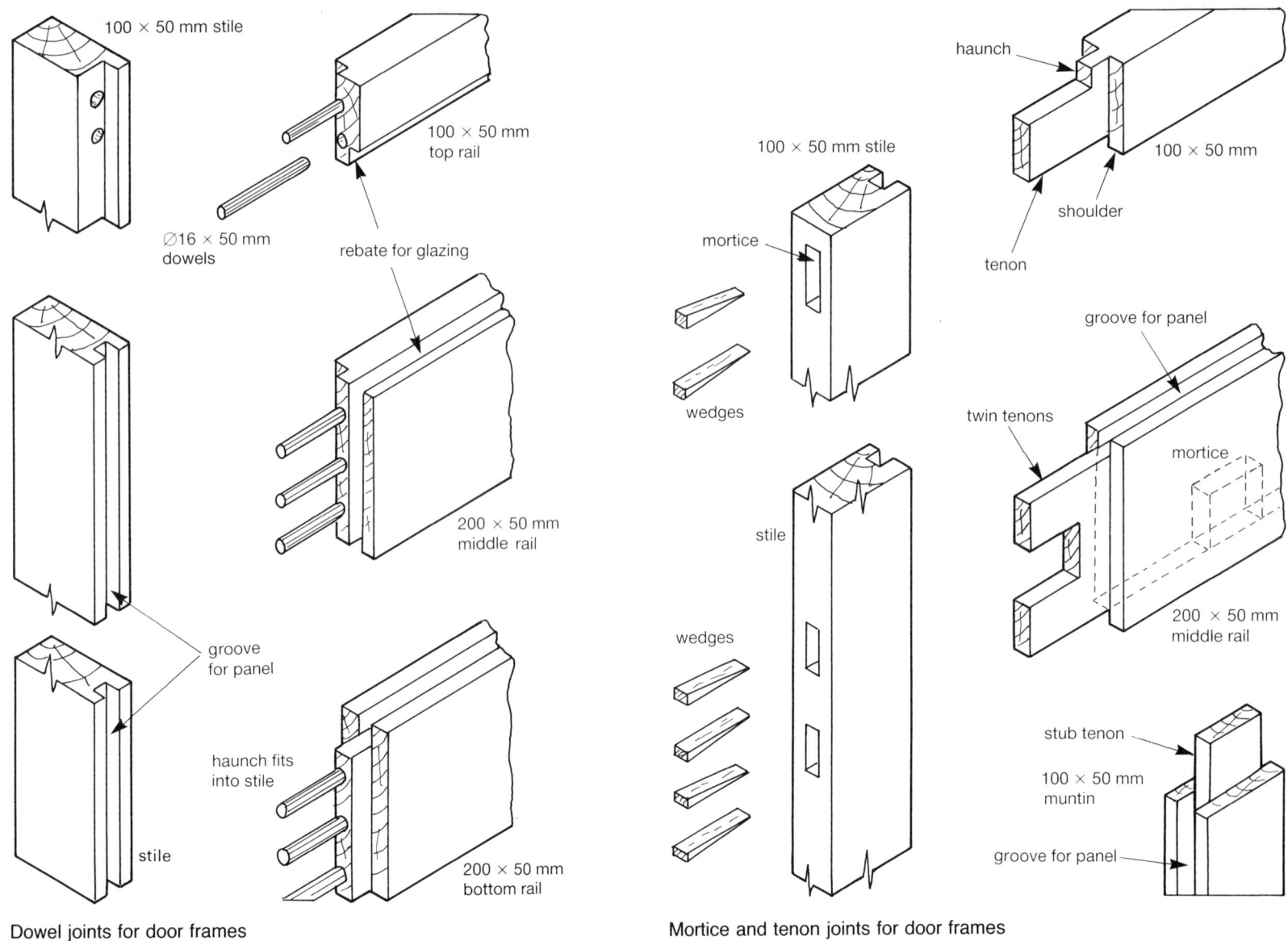

Mortice and tenon joints for door frames

DOORS AND WINDOWS

Flush doors

Flush doors are now very popular, on account of their plainness, low cost and ease of construction. Flush doors are available in the standard sizes indicated earlier and their thicknesses vary from 35 mm to 44 mm. A flush door consists of a core of either laminated solid timber or skeleton framework with a facing of plywood.

The usual forms of construction are shown below. The strongest type of flush door is the solid core, which is constructed using strips of timber that are laminated to form a solid board, with a facing of plywood on both sides.

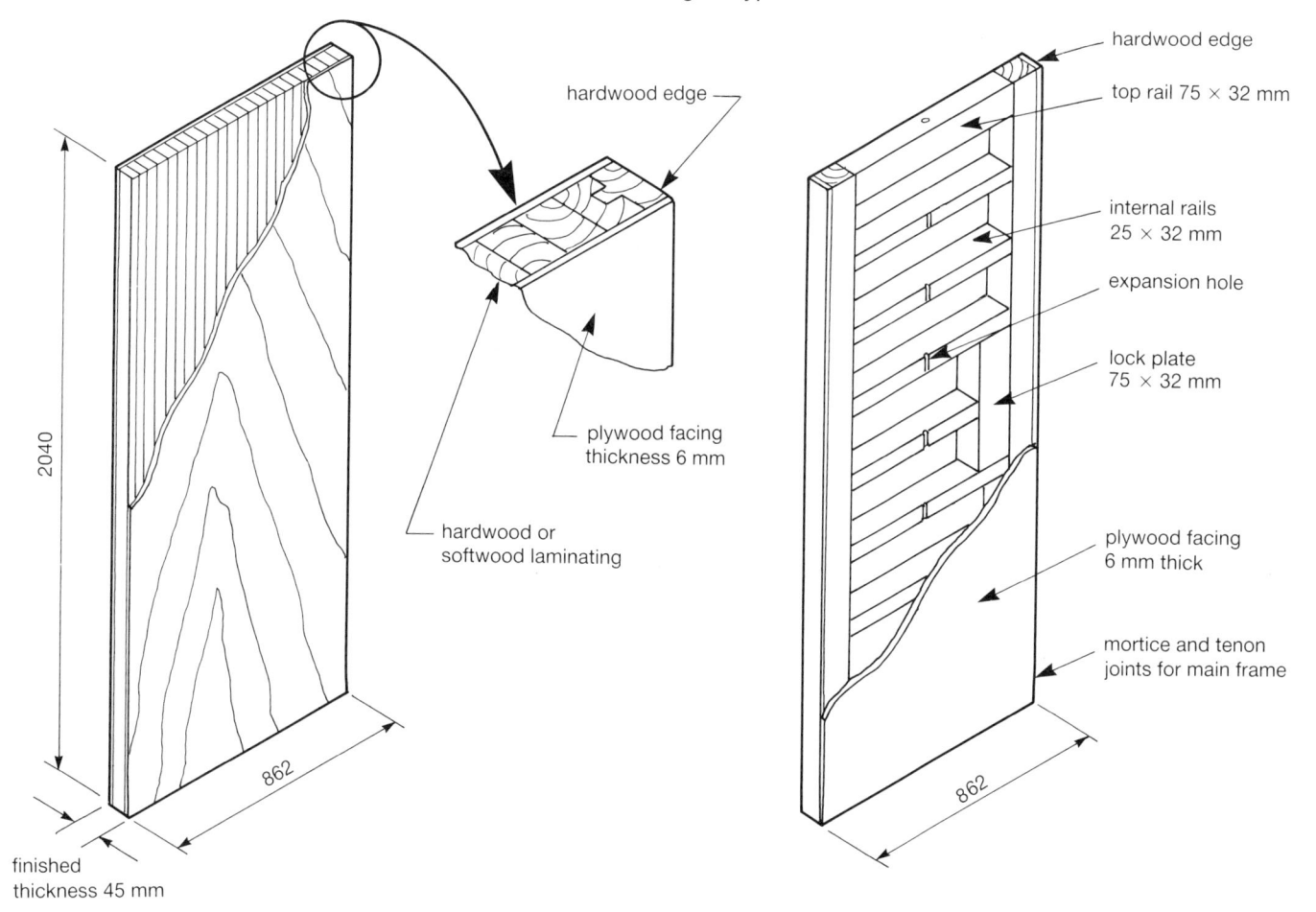

Laminated solid timber flush door

Skeleton framework (half solid)

51

Match-boarded doors

Match-boarded doors are mostly used for gates and as garage and store doors. A match-boarded door may be ledged, or it may be framed, ledged and braced. A framed and ledged match-boarded door is shown below. Match-boarded doors are sometimes also called braced and battened doors. The top, middle and bottom rails constitute the ledges.

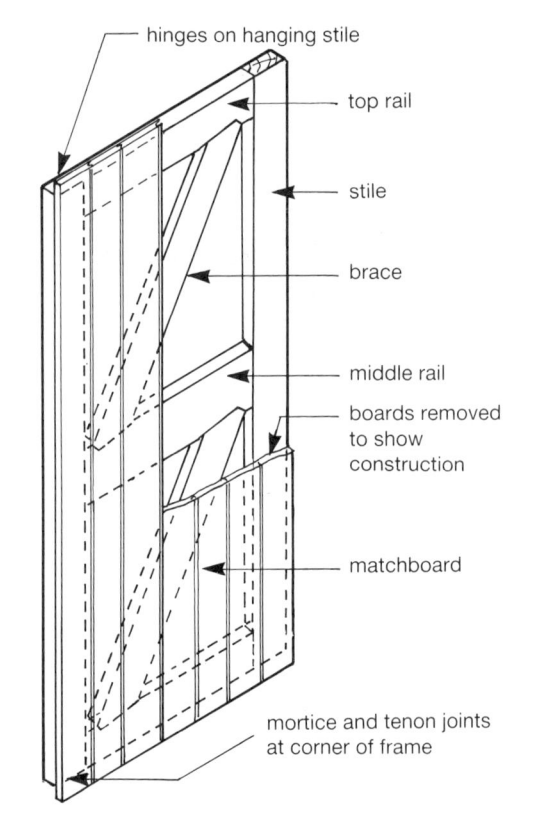

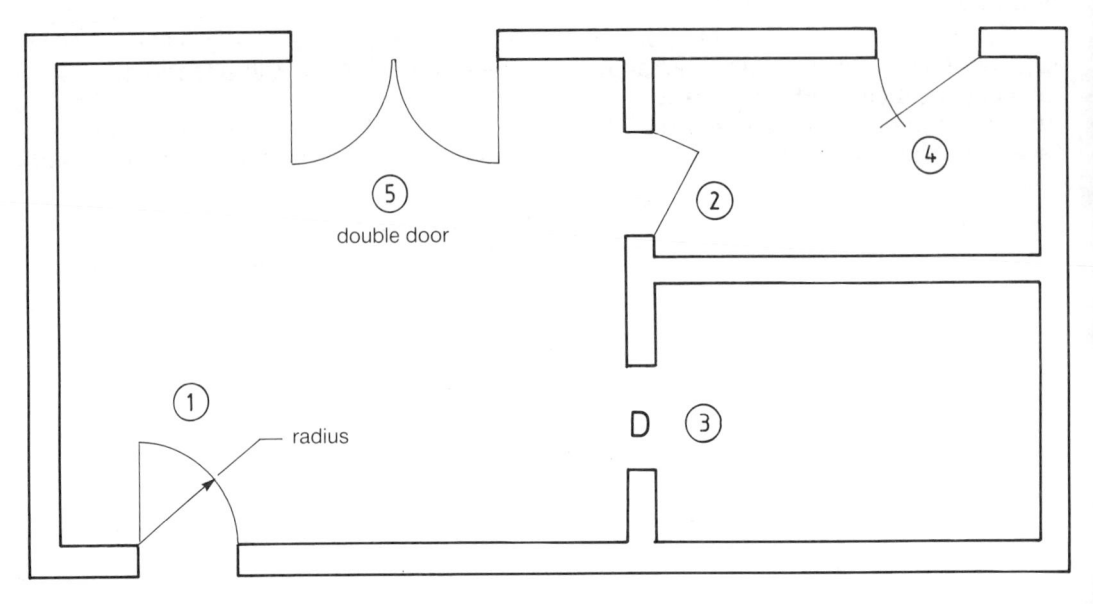

Different methods of representing doors

Framed, ledged and braced door

Representing doors on plans

Although there are various types of doors, the methods of representing them in floor plans are the same, and are relatively simple. The drawing above shows the most common methods. In all the methods shown, it will be noticed that essentially a door is represented by a wall opening and a symbol showing the direction of opening or swing of the door.

Method 1 involves using the width of the door opening as the radius for describing an arc whose centre is the corner of the opening on which the door will be hinged. A line at right angles to the wall determines the terminal point of the door swing.

Method 2 consists of drawing two lines which form an angle and which span the door opening. The longer of these lines is of length equal to the width of the door opening and is located on the side to which the door will swing.

Method 3 simply shows the door opening, which is sometimes accompanied by the capital letter D to indicate that a door is located within the distance shown.

DOORS AND WINDOWS

Method 4 is only a variation of method 2. The difference between the two is that the arc of the former is shorter and the straight line joining the terminal point of the arc to its centre is not at right angles to the wall but inclined at about 30° to the wall. The centre of the circle of which the arc is a part is located at the opposite side of the door opening.

A double door such as those used for garages or main entrances to business premises is represented by a double swing (method 5).

Determining the width of door openings

Usually, doors (flush or panelled) for domestic buildings are built to standard sizes, the most common being 750 mm wide. The frames for most doors are usually about 75 mm thick with rebates that cover the edges of the door.

The external or out-to-out width of the frame is 900 mm, which is the width of the door opening. In a drawing, therefore, the width of the door is located at the desired positions on the floor plan. When locating doors on the floor plan it is important to avoid putting them so close together that the opening of one hinders traffic through, or interferes with the opening of, another door. For example, if door 5 were to be moved towards door 2 in the plan on page 52, there would be interference in the opening and closing of one or both doors.

Representing doors in elevation

Unlike in the floor plan, where all types of doors are represented in the same way, in elevations doors are drawn the way they actually look.

On elevations doors are usually located by projecting their locations from the floor plans to the elevations. The picture below shows how the doors on the front elevation of a building are located by projecting from the floor plan. The height of the door is usually about 2100 mm and is measured from the floor level of the building.

Observe that single lines marking the edges of the door opening are projected to the elevation. To make the doors in the elevation appear real to the eyes, double lines are drawn to represent the door frames. Also, a door handle and sometimes a lock are added since these are usually seen.

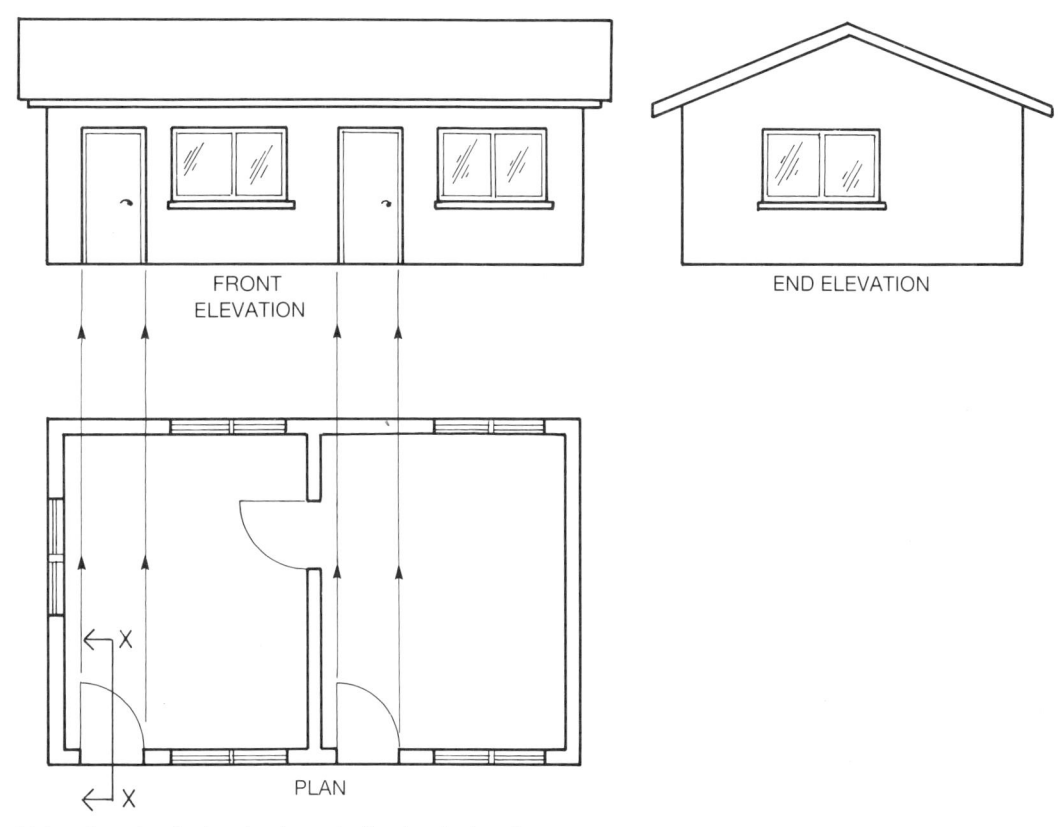

Using the plan to locate doors in the front elevation

TECHNICAL DRAWING 3: BUILDING DRAWING

Drawing a door in section

If the cutting plane for a section passes vertically through a door, then that door must be shown in section. Sometimes it is necessary to indicate all the construction details of the door in the section. It is good practice also to show these details separately.

A section X-X through the internal door of the building on page 53 is shown on the right. When this section is visualised the following features will be observed:
1. the lintel and the line of contact between the lintel and the head of the door frame,
2. the head of the door frame,
3. an outer vertical line a) representing the wall and an inner vertical line d) representing the edge of the door,
4. two middle vertical lines b) and c), where b) represents the edge of the frame stile and c) represents the inner edge of the door,
5. the core of the flush door or details of any other type of door.

Thus, the most common features usually shown in sectional drawings of doors are the head of the door frame, the frame stile and construction details of the door. Note that the width of the head of the frame is about 75 mm and the thickness of the flush door is 40 mm.

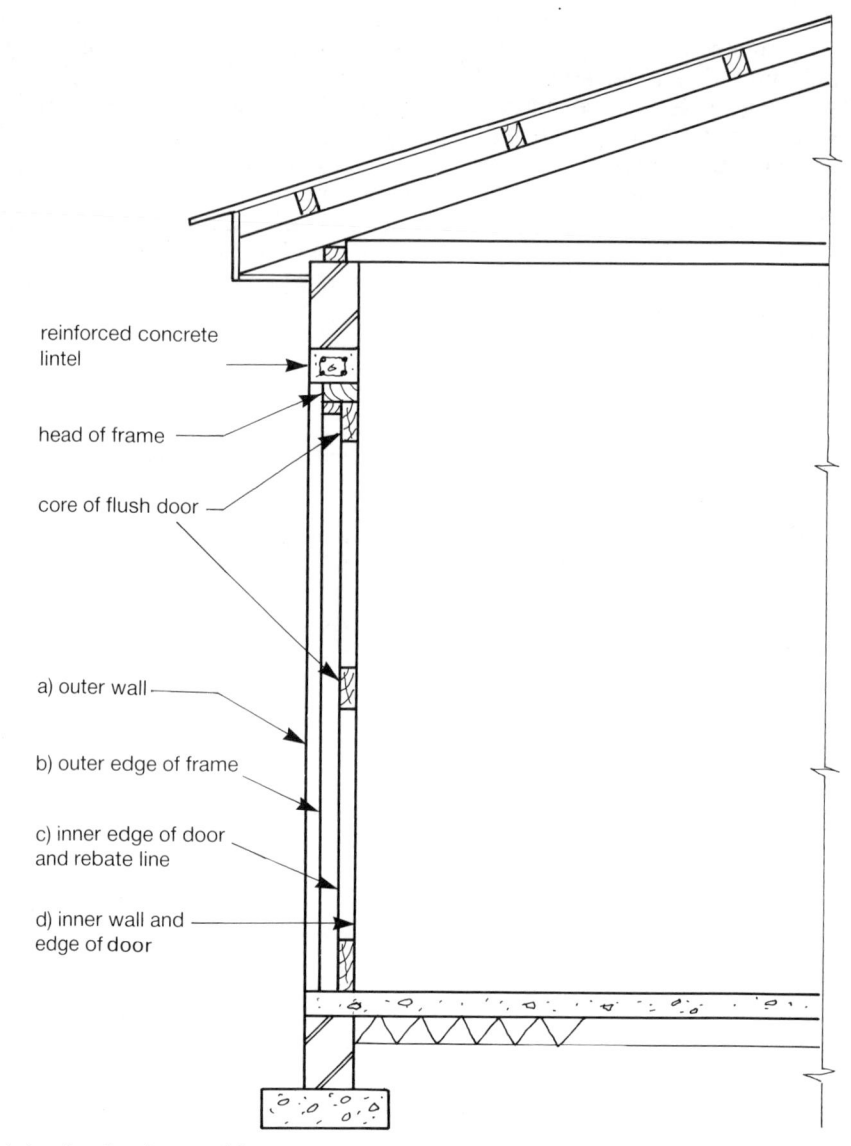

Sectional details of a door and frame

Windows

Windows may be divided into classes according to the manner of opening them, namely:
1. casement windows: these are hung on the side or top using hinges,
2. double-hung sash windows: these slide vertically,
3. pivoted sash windows: these are hung either horizontally or vertically,
4. sliding sashes which slide horizontally,
5. louvres in which the blades open or close simultaneously on being actuated by a lever.

Windows may also be classified by the material from which they are made, such as steel, aluminium and timber.

Casement and louvre windows are more commonly used in residential building construction than the other types of windows. Apart from the louvre type of window, they all look alike. The drawing above right shows the elevation, plan and section of a casement window. The drawing below right shows the view of a louvre window.

Window and frame sizes

Window frames for louvres are usually made with timber of about 100 mm by 50 mm thickness. Frames for casement windows are usually thicker: 100 mm x 75 mm and are rebated to accommodate the sashes. Standard casement windows are made in widths of 438 mm, 641 mm, 1226 mm, 1810 mm, 2394 mm and heights of 768 mm, 1073 mm, 1226 mm, 1378 mm and 1530 mm. Each width of a casement window can be subdivided into lights.

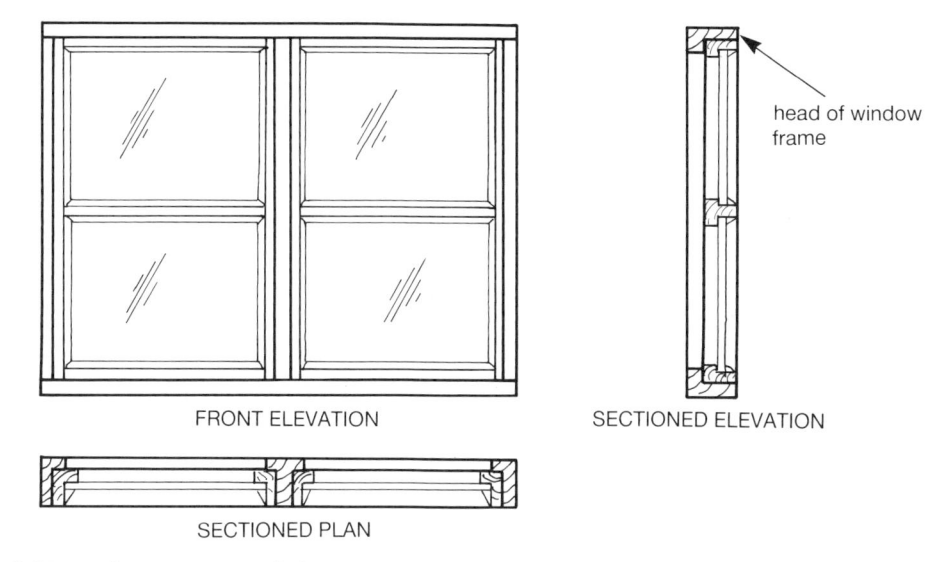

FRONT ELEVATION

SECTIONED PLAN

Two light wooden casement window

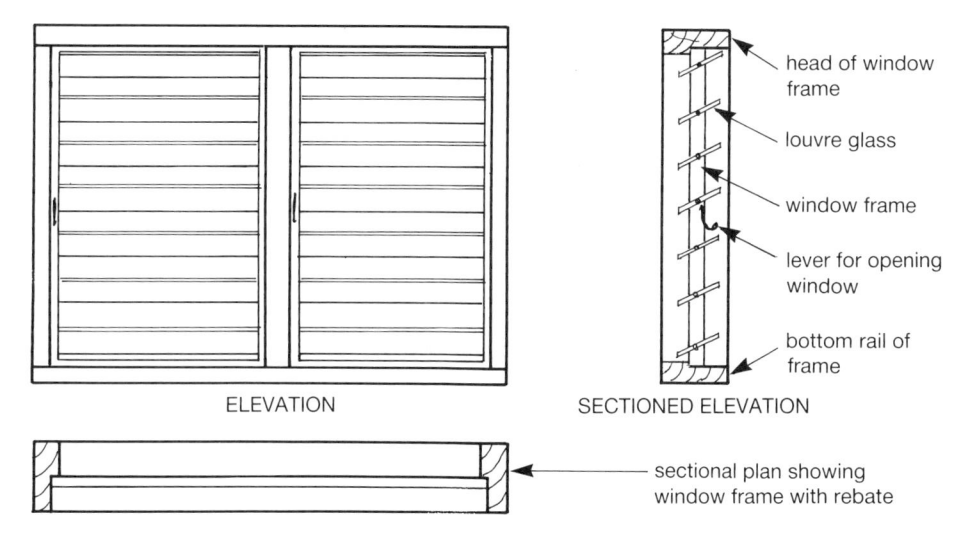

SECTIONED ELEVATION — head of window frame

ELEVATION

Two bay louvre window

SECTIONED ELEVATION — head of window frame, louvre glass, window frame, lever for opening window, bottom rail of frame

sectional plan showing window frame with rebate

Standard louvre blades and frames are available in spans ranging from 450 mm to 750 mm and heights which depend on the number of louvre blades desired. Since most blades are 150 mm wide, a window 1200 mm high may require about eight blades.

Locating windows in walls

There are two important considerations. Firstly, a window should be located so as to be central in relation to the width or length of the room except where it is impossible to do so, such as in cases where there are doors or wardrobes. In the top drawing, the windows W_3 and W_6 are not located centrally in the rooms, whereas windows W_1, W_2, W_4 and W_5 are centrally located.

The second consideration relates to the representation of the window frame in the wall thickness. Most wall thicknesses are 150 mm or 225 mm. In either case, the thickness of the wall is wider than the width of the timber frame. The window frame, both in construction and in drawing, is positioned so that, in the case of a 150 mm-thick wall and a 100 mm frame, the excess 50 mm on the wall is left on the inside rather than on the outside. The bottom lefthand drawing shows a 100 mm by 50 mm window frame located in a 150 mm wall. Sometimes the frame is centred in the wall (right).

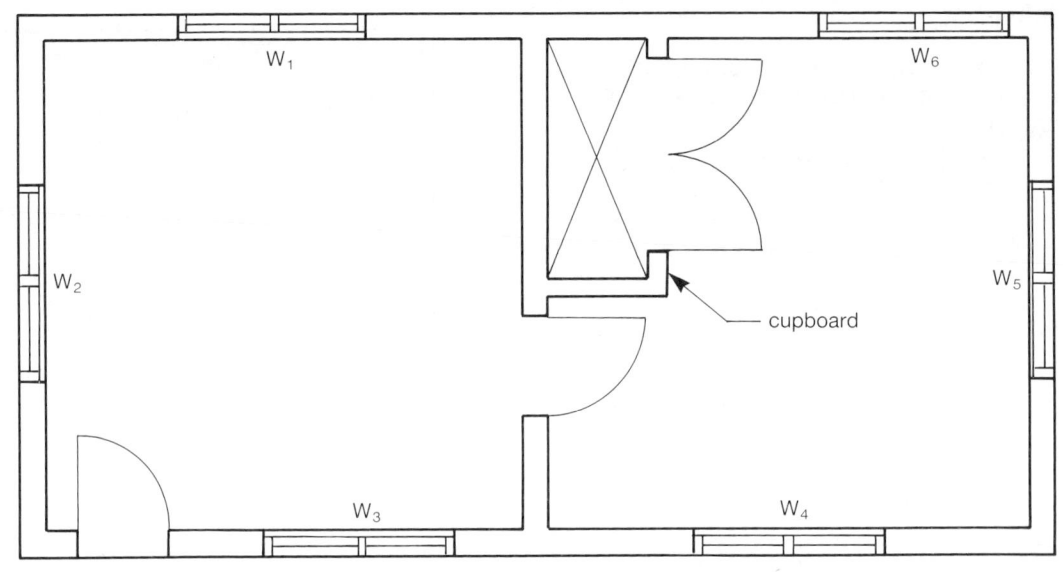

Locating windows in a building plan

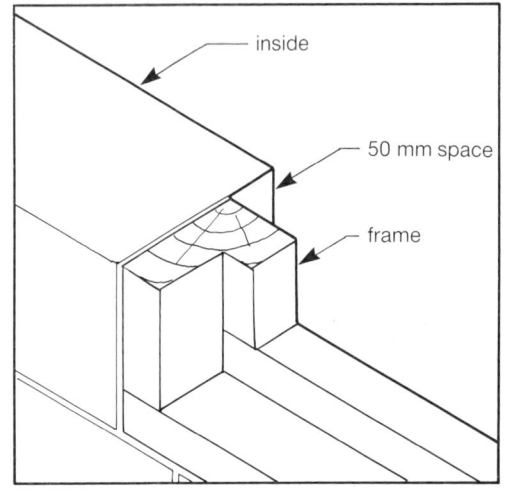

Window flush with outside of wall

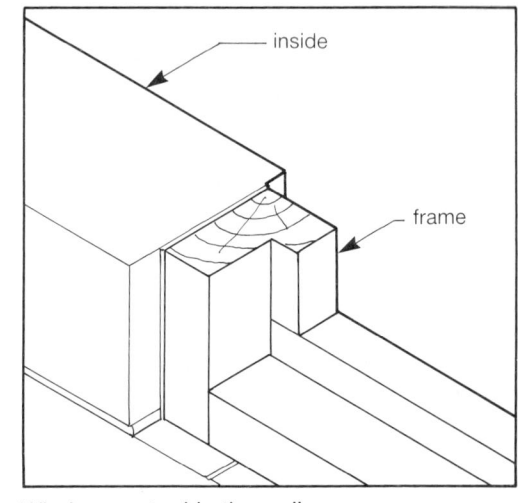

Window centred in the wall

DOORS AND WINDOWS

In drawing, the window frame will appear on the plan as shown on the right. The main features in the plan of a window when the louvre glasses have been fitted are shown.

If the frame is as wide as the thickness of the wall only four horizontal lines will be seen and not five. The following points should be noted as guides in drawing windows:

1. The width of the window opening (W) is determined by the size of the window to be installed. For example, if a two-bay louvre window is to be installed, the width of each bay (or length of the glass blades) is 600 mm; if the thickness of the frame is 50 mm, then W will be equal to (50 + 600 + 50 + 600 + 50) mm = 1350 mm. Therefore, to draw the window on the plan, the distance to be marked on the drawing sheet to represent the window opening is 1350 mm.

2. In order to draw the windows in elevation, projection lines are drawn from the plan to meet the front elevation of the building. Windows which are in the front wall are then projected to the front elevation, for purposes of determining the width of such windows in the elevation (see page 58).

3. The height of the window is determined as follows:
   a) Measure a height of 2100 mm from the ground level of the elevation if the ground level corresponds to the floor level. If the ground level does not correspond to the finished floor level, measure the height given above from

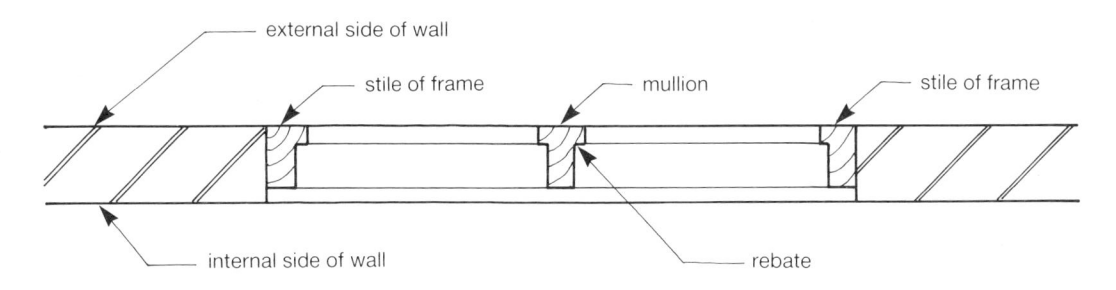

Plan of a rebated frame in a wall

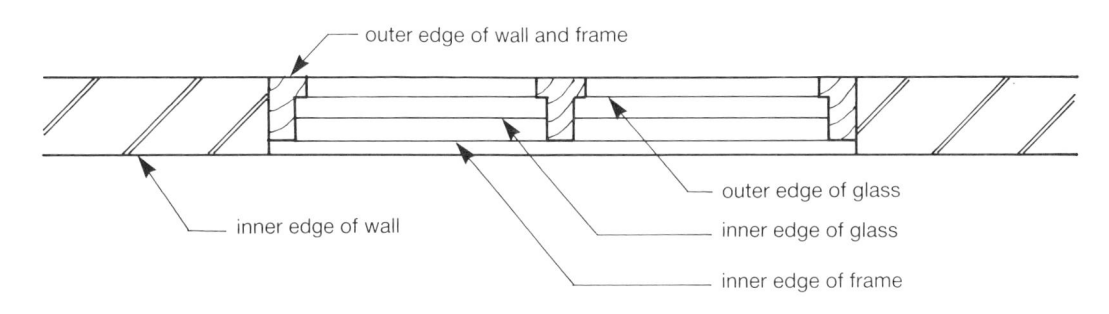

Plan of a louvre window in the frame

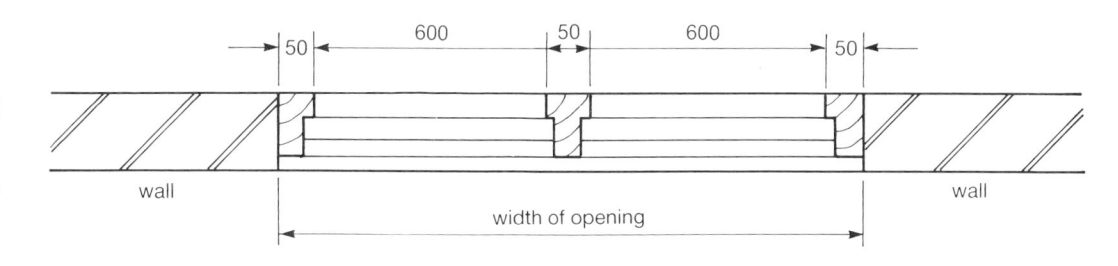

Dimensions of the window

57

ELEVATION

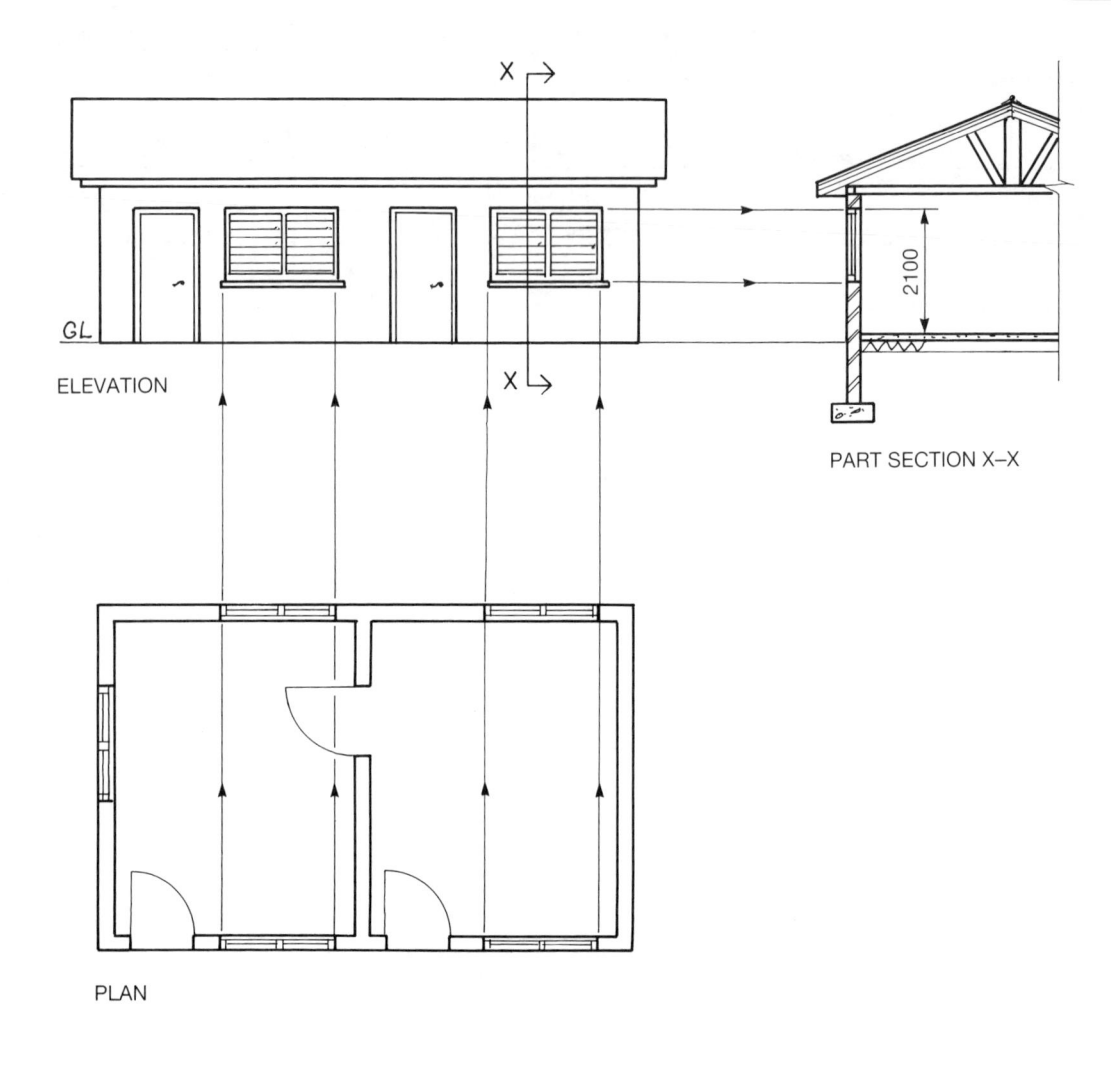

PART SECTION X–X

PLAN

Drawing windows in the elevation

the level of the finished floor. This height represents the top most part of the window.

b) Measure the actual height of the window from the 2100 mm mark downwards. For example, if the window is a 2-bay, 8-blade louvre type, the height should be approximately 1200 mm. To calculate the exact height of the window in this example, note that each blade is 150 mm wide. However, when the windows are fixed, there is usually an overlap of about 12.5 mm. For example, for eight blades there are seven overlaps, which give a total of 87.5 mm. In effect, the eight blades will occupy a space of about [(150 × 8) − 87.5] mm = 1112.5 mm. Assuming the frame to be 50 mm thick, the head of the frame and the bottom rail will give 100 mm. Also, the lowest blade usually goes into a rebate up to 12.5 mm. Therefore, the total height of the window opening will be [(1112.5 + 100) − 12.5] mm = 1200 mm. The above calculation is shown pictorially on page 59.

c) Once the height of the window opening has been determined, mark out the thickness of the frame head and bottom rail from the overall height obtained.

d) The space for the louvres is then divided equally according to the number of louvre blades but with the lowest blade being 12.5 mm less than the others since this distance is usually covered by the rebate or not readily visible.

The drawing of the window when finished should look as shown below. Note that the dimensions are not indicated in actual drawing practice but have been included in this illustration for purposes of clarity.

If there is a window to be shown in the end elevation, the procedure just outlined for the front elevation should be followed. It is important, however, that the windows should be projected from the appropriate ends on to the plan.

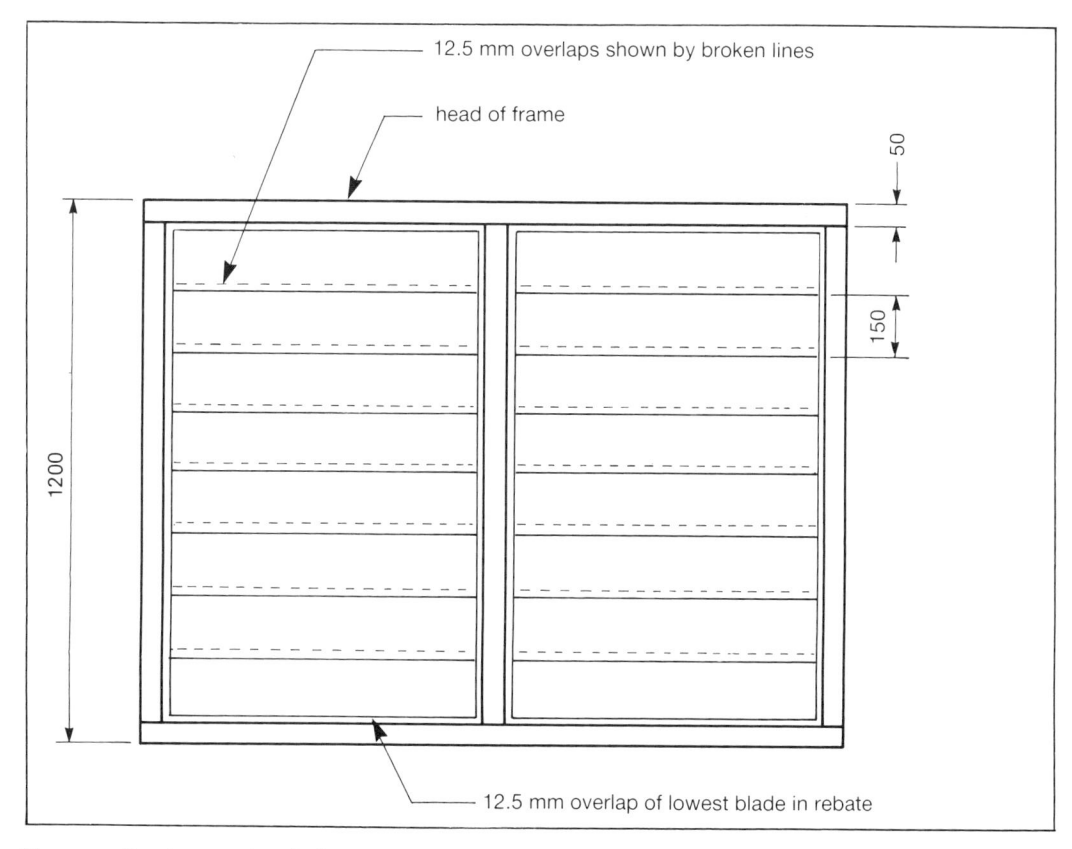

Drawing windows in sections

The drawing on page 60 shows a section through a louvre window. When a window is sectioned, the features to be seen include the lintel and the head of the frame, at the upper level. At the lower level the bottom rail of the frame will also be seen. Between these levels there are five (and sometimes four, if the frame is as wide as the wall) vertical lines. See the key to the vertical lines in the same drawing.

To draw a window in section, the height should be measured from the floor level as earlier described for the front elevation. Thereafter:
1. mark out the frame head and bottom rail,
2. mark out the vertical lines between the two sides of the wall to represent the frame for louvres and the inner edge of the frame stile (if the wall is wider than the frame stile). The frame is usually placed in the middle of the frame stile.

Representing louvres in windows

DOORS AND WINDOWS

59

TECHNICAL DRAWING 3: BUILDING DRAWING

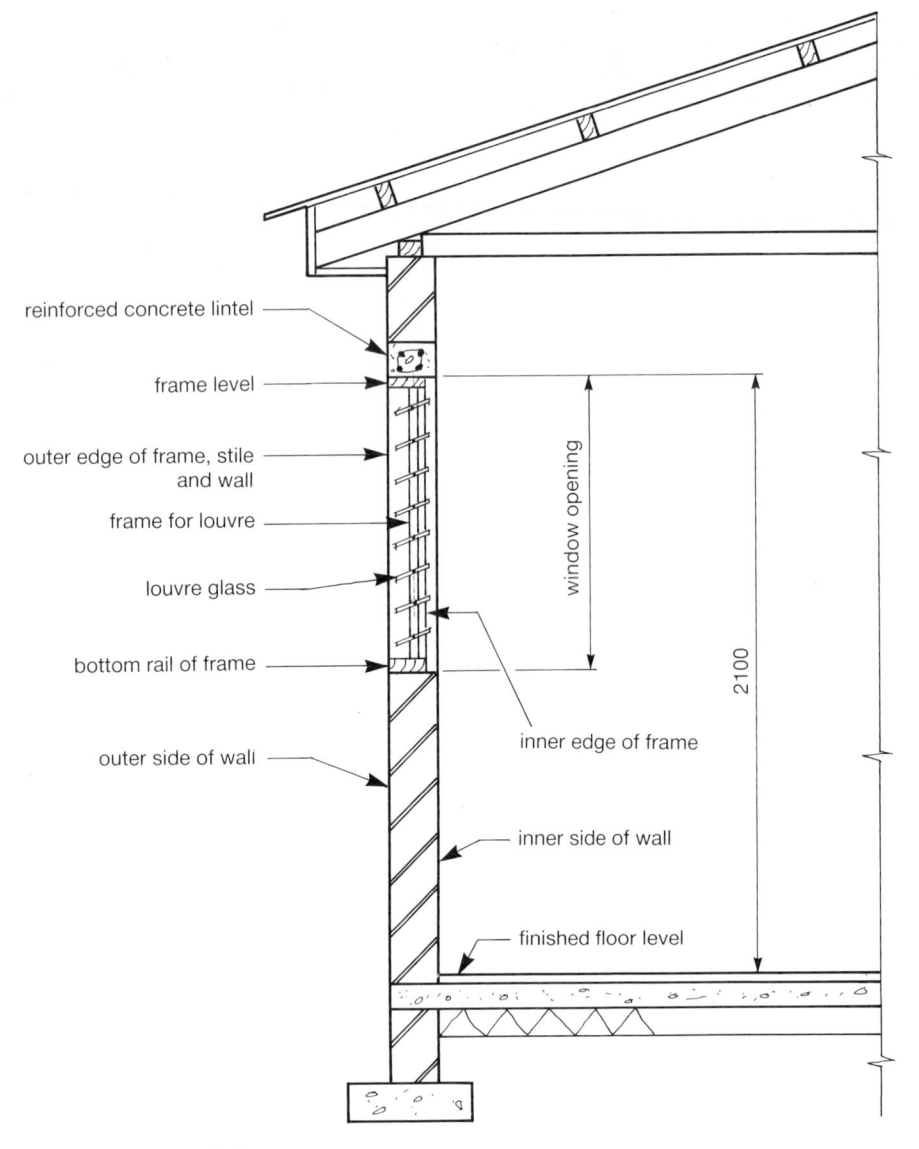

Exercises

1. Draw a section of a three-panelled door and label all its parts.
2. Using appropriate sketches, illustrate each of the following:
   a) mortice and tenon joint,
   b) dowel joint,
   c) tongue and groove joint.
3. By means of a neat sketch, show the construction details of a half-solid flush door.
4. Draw a section through the plan of a two-room house and show correct methods of representing doors and windows in section.
5. Make good sketches to illustrate the following:
   a) plan of a window fitted with louvre frames and 10 blades,
   b) the front elevation of a window fitted with louvre frames and blades,
   c) a section through the window you have drawn in b).

Section through a louvre window

60

9. Stairs

Stairs consist of a series of steps with accompanying handrails which provide easy access to various floors or levels of a building. Two important considerations in the design of a stair are easy ascent or descent and safety. The steps of a stair may be constructed of either concrete or timber; the rails may be of timber or metal.

Types of stairs

Straight-flight stair

This is the simplest form of stair and consists of a straight flight or run of parallel steps (see right). As the name suggests, a straight-flight stair has no turns, but it may have a landing between flights.

L-stair

The L-stair (see right) has one landing at some point along a flight of steps. If one arm of the L-shape is longer than the other, that is if the landing is nearer the top or bottom of the stair, the stair is referred to as a long L-stair. L-stairs are used when the space required for a straight-flight stair is not available.

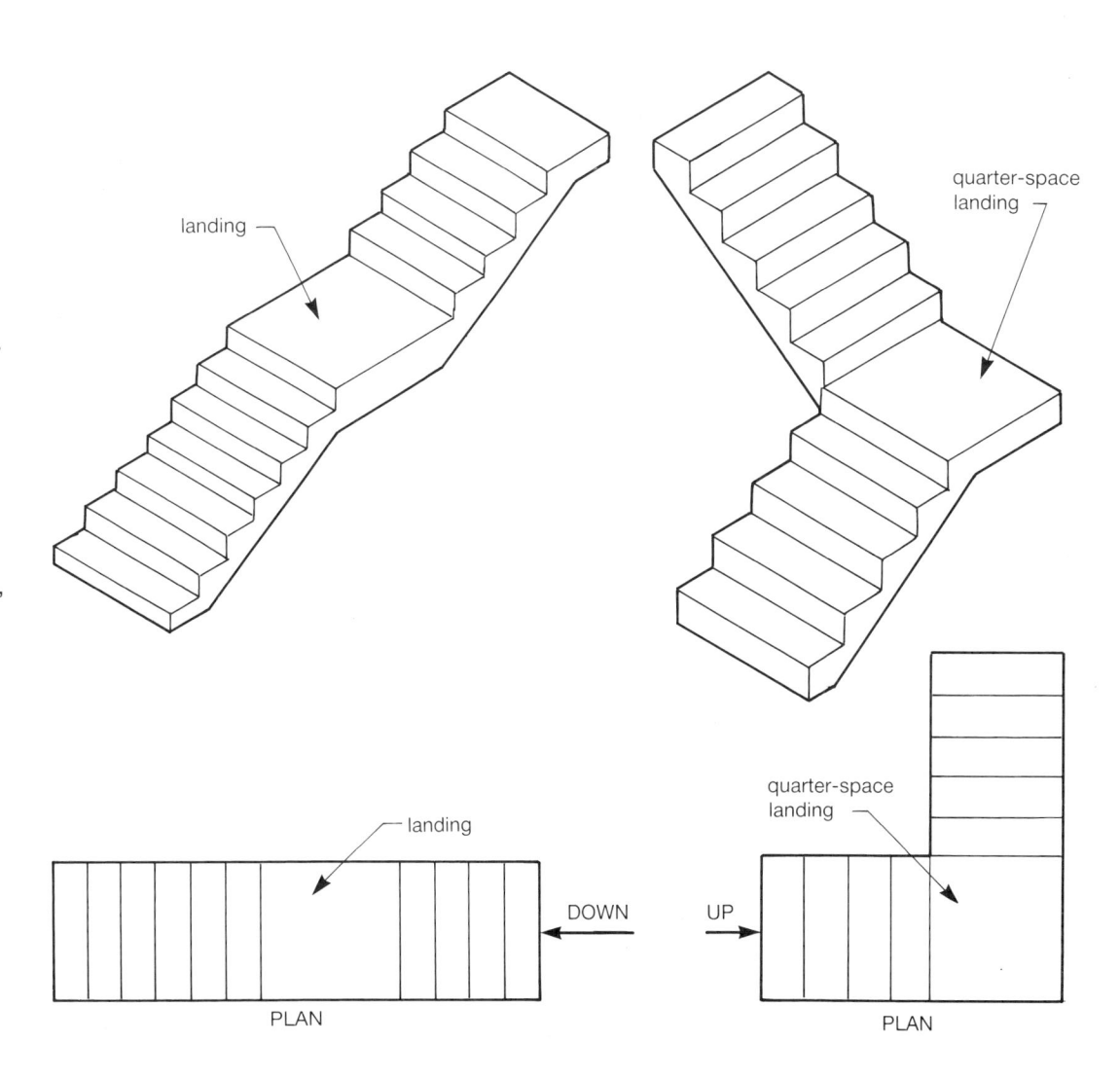

61

TECHNICAL DRAWING 3: BUILDING DRAWING

Dog-leg stair

A dog-leg stair (see right), sometimes called a half-turn stair, has one flight rising to an intermediate half-space landing, with the second flight running in the opposite direction to the first flight and parallel to it.

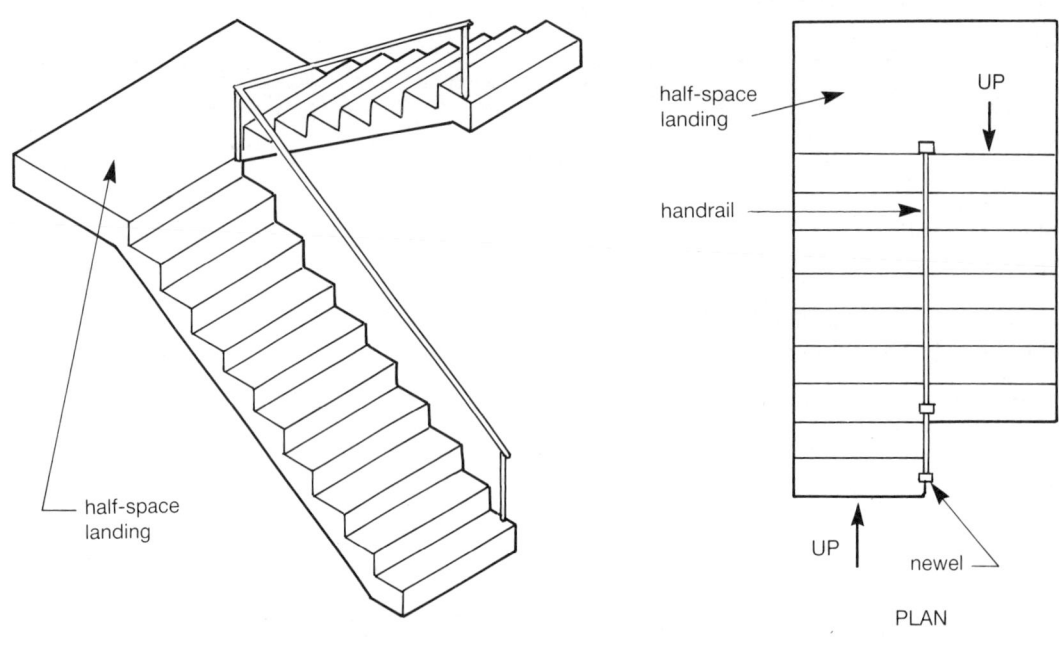

Open-well or open-newel stair

This stair (see right) has a central well hole between its two parallel flights. Newels are located at each change of direction of flight.

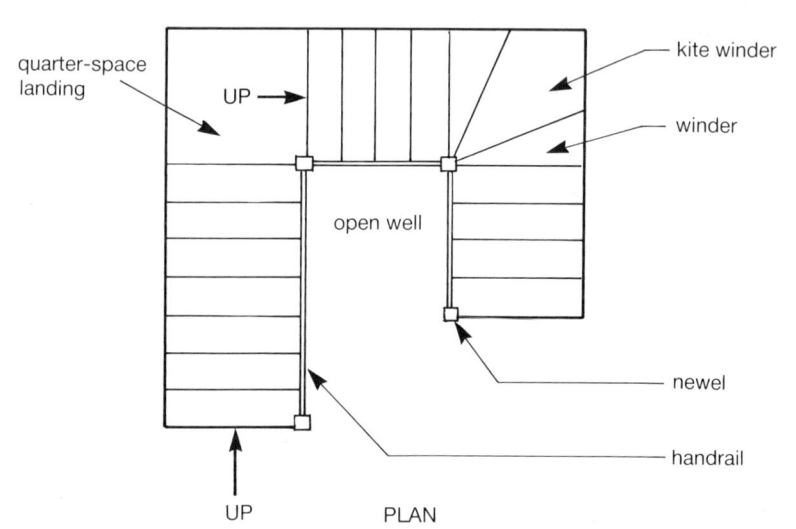

STAIRS

Geometrical stair

A geometrical stair (see below) is in the form of a spiral, with the face of the steps radiating from the centre of a circle which forms the plan of the outer string. A geometrical stair has an open well similar to that of an open-well stair. A spiral stair is a form of geometrical stair but has no well. It may be used where little space is available.

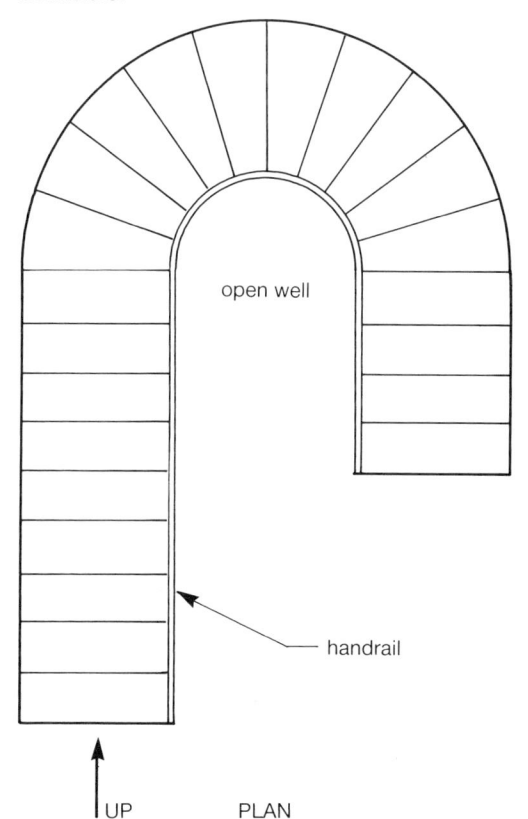

PLAN

Stair design

Terms used in connection with stairs

For a better understanding of stair design it may be necessary here to look at some of the technical terms associated with stairs. Some of the important terms include:

1. *Tread*: The horizontal member of each step.
2. *Riser*: The vertical face or member of each step.
3. *Step*: The combination of tread and riser.
4. *Nosing*: The rounded projection of the tread which extends past the face of the riser.
5. *Rise*: The vertical distance between two consecutive treads.
6. *Going or run*: The horizontal distance between the nosing of a tread and the nosing of the tread or landing next above it.
7. *String, stringer or carriage*: A structural member which supports the treads and risers.

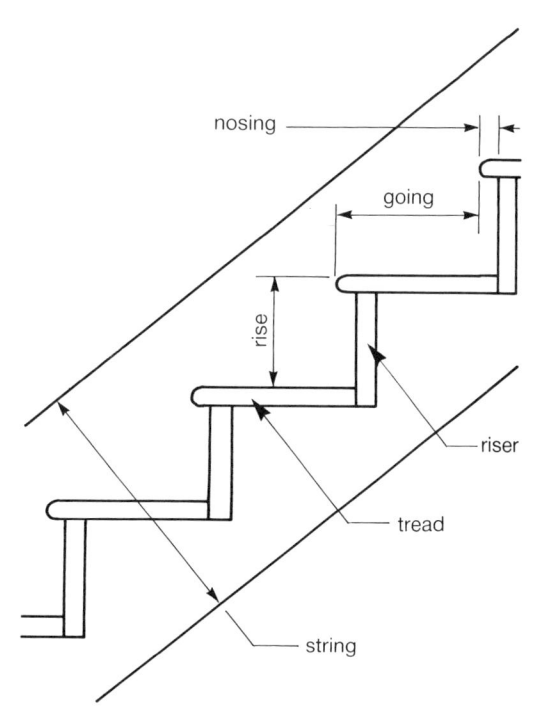

Parts of a wooden stair

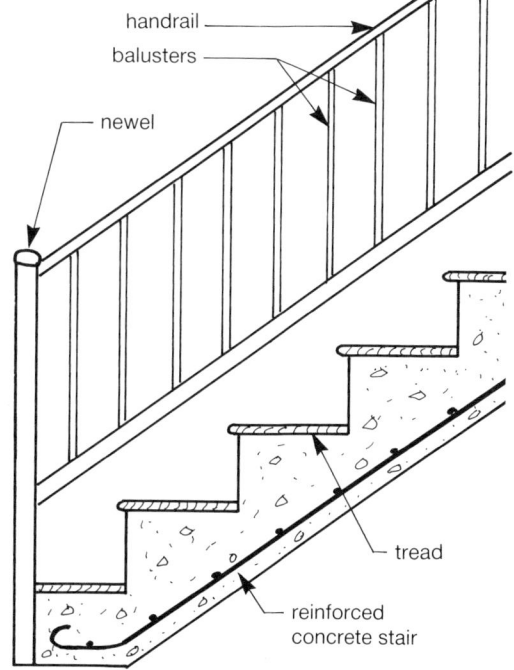

Balustrade

63

8. *Winder*: A tapering step where the stair changes direction, radiating from a newel.
9. *Newel*: The post at the end of a flight to which the stringers and handrail are fixed.
10. *Balusters*: Vertical members which support the handrail.
11. *Balustrade*: A framework of handrail and balusters.
12. *Landing*: The floor area at some point between or at either end of a flight of stairs.
13. *Headroom*: The shortest clear vertical distance measured between the nosing of the treads and the floor immediately above or the ceiling (see right).
14. *Stairwell or staircase*: The opening in which a set of stairs are constructed.

Drawing of stair details

The drawing showing stair details is an important one among a set of working drawings for storey buildings or for buildings having more than one floor level. Stair details show the structure, stair size, posts, rails and balusters, materials and method of attachment to the structure.

Points to note when designing a stair

1. The ascent of the stair should be relatively easy. The stair should not be too steep, nor should the incline be at too small an angle. In fact the pitch of a domestic private stair should not exceed 42° and that of any other building not more than 38°.

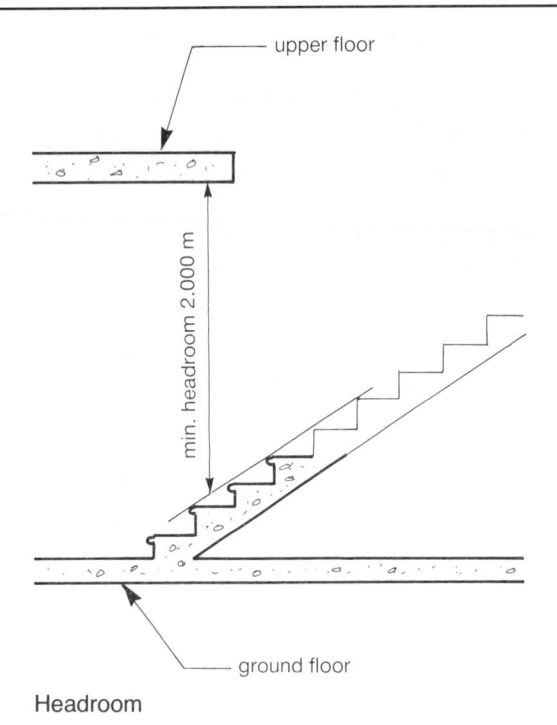

Headroom

2. All the risers must be of the same height, and the treads should be of uniform width in order to avoid accidents. An ideal riser height is 187 mm and an ideal going is 265 mm.
3. The maximum number of steps in a flight is preferably 12, but for common stairs (say in public buildings), the maximum should be 16.
4. The width of the stair should not be less than 900 mm from wall to wall, or wall to centre of outer string.
5. Adequate headroom must be provided and this should not be less than 2 m.
6. The height of the handrails from the nosing line should not be less than 840 mm.

Procedure for drawing a stair in section

1. Choose a convenient scale. A scale of 1:50 may be suitable.
2. Determine and draw the distance from the finished ground floor level to the finished upper floor level. This is the total rise of the stair. For a concrete upper floor whose ceiling height from the finished ground floor is, say, 3 m and with a floor slab thickness of 150 mm, the total rise will be the sum of the ceiling height and the thickness of the floor, i.e. 3.000 m + 0.150 m = 3.150 m.
3. Determine and lay out the total going. This is usually specified in the instruction given regarding the drawing. For example assume that the total going is 3.600 m (see below).

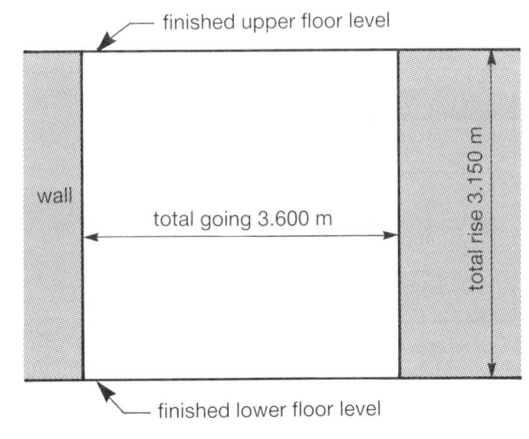

Layout of floor levels

4. Calculate the tread and riser dimensions as follows:

Each rise is calculated from the formula,

$$\text{rise} = \frac{\text{total rise}}{\text{number of risers}}$$

In the example above, the total rise is 3.150 m. To determine the number of risers involves some degree of trial and error, but each trial is based on what should be an ideal riser height (187 mm), and the formula used is going + twice rise $(G + 2R) = 550$ mm to 700 mm. If we divide 3.150 m by 187 mm we shall have 16.3 risers. Since we cannot have a fraction of a step we shall have either 16 or 17 risers. Assuming 17 risers, then each rise

$$= \frac{3.150 \text{ m}}{17} = 185 \text{ mm}.$$

This is close to an ideal riser height (or rise). Each going is calculated as follows:

$$\text{going} = \frac{\text{total going}}{\text{number of treads}}$$

We are given the total going as 3.6 m. Since the number of treads is one less than the number of risers, we shall have 16 treads in all. Therefore each going

$$= \frac{3.600 \text{ m}}{16} = 225 \text{ mm}.$$

The values of the rise and the going are checked for suitability using the formula

$$G + 2R = 550 \text{ to } 700 \text{ mm}.$$

Substituting gives $225 + 370 = 595$ mm.

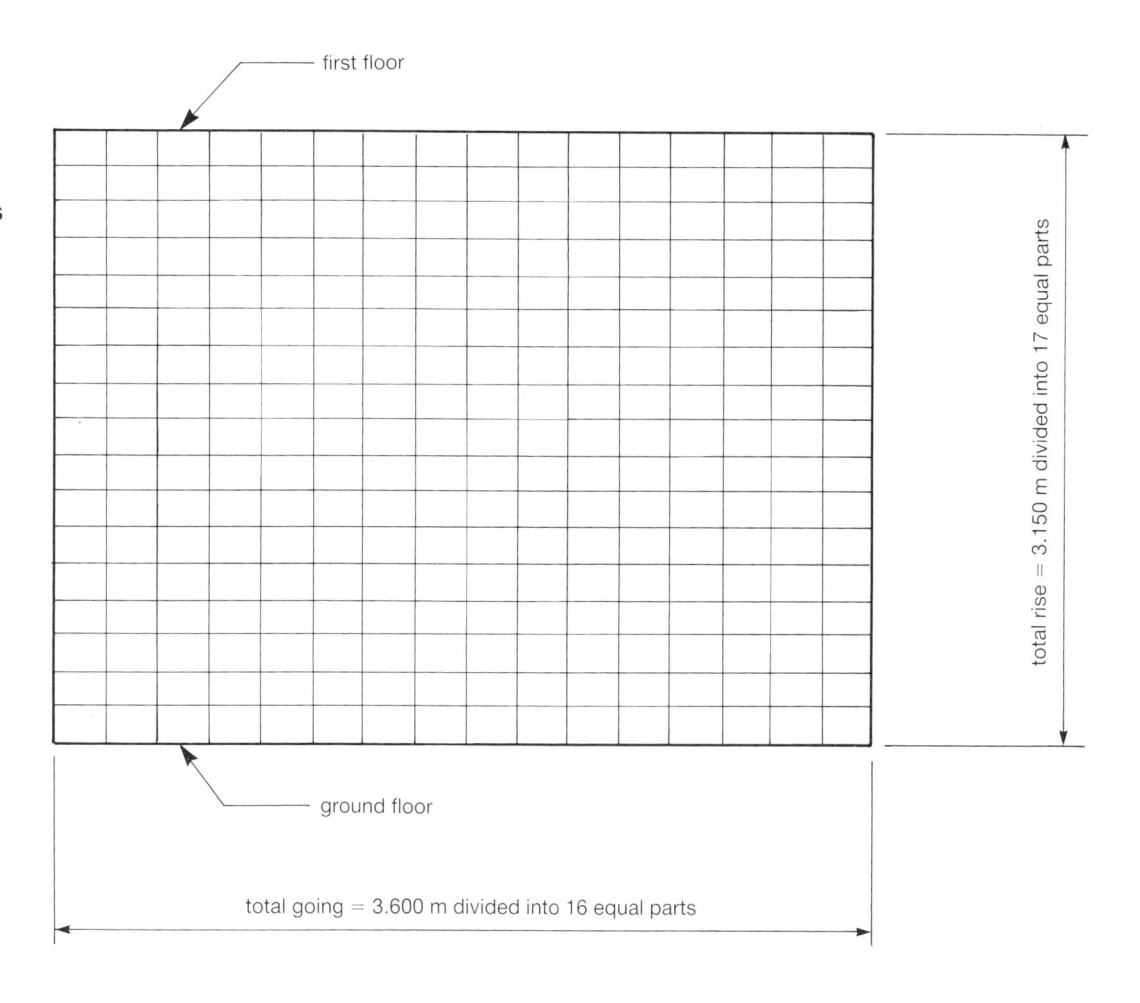

Dividing the total going and the total rise into the number of risers and treads required

Since 595 mm falls within the range 550 to 700 mm it is considered suitable.

With the data obtained from our previous calculation, namely:

total rise	= 3.150 m
total going	= 3.600 m
rise	= 185 mm
going	= 225 mm

divide the total rise and the total going or run into the number of risers and treads required (see drawing on page 65) if you are drawing a straight-flight stair. If a newel stair is being drawn, appropriate landings must be included in the design, and the flight will not be in one direction.

5. Form the steps by darkening the tread and riser profiles. Draw the bottom edge of the concrete (see right). This should be a minimum of 100 mm from the line joining the underside of the steps as shown.
6. Locate the stairwell through opening, bearing in mind that the headroom dimension will affect the size of the stairwell through opening. Remember that the minimum headroom is 2 m.
7. Locate the newel post on the first step on the ground floor and on the landing or top floor as shown on the left. If we assume a diameter or square cross-section of side 75 mm for the newel post, this dimension should be located in the centre of the width of the tread as shown.

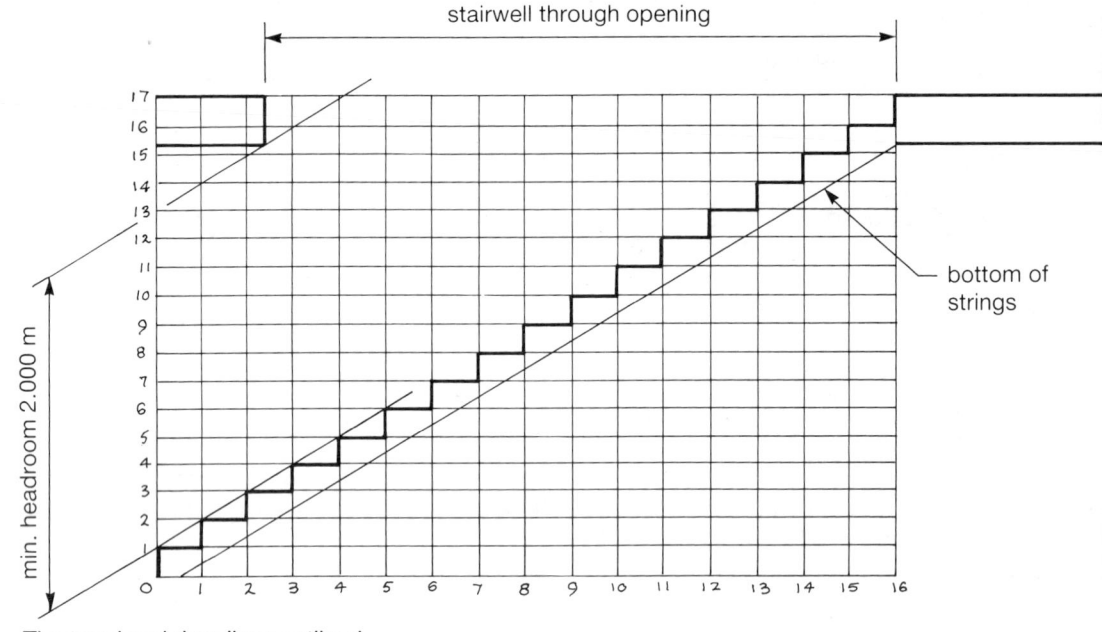

The tread and riser lines outlined

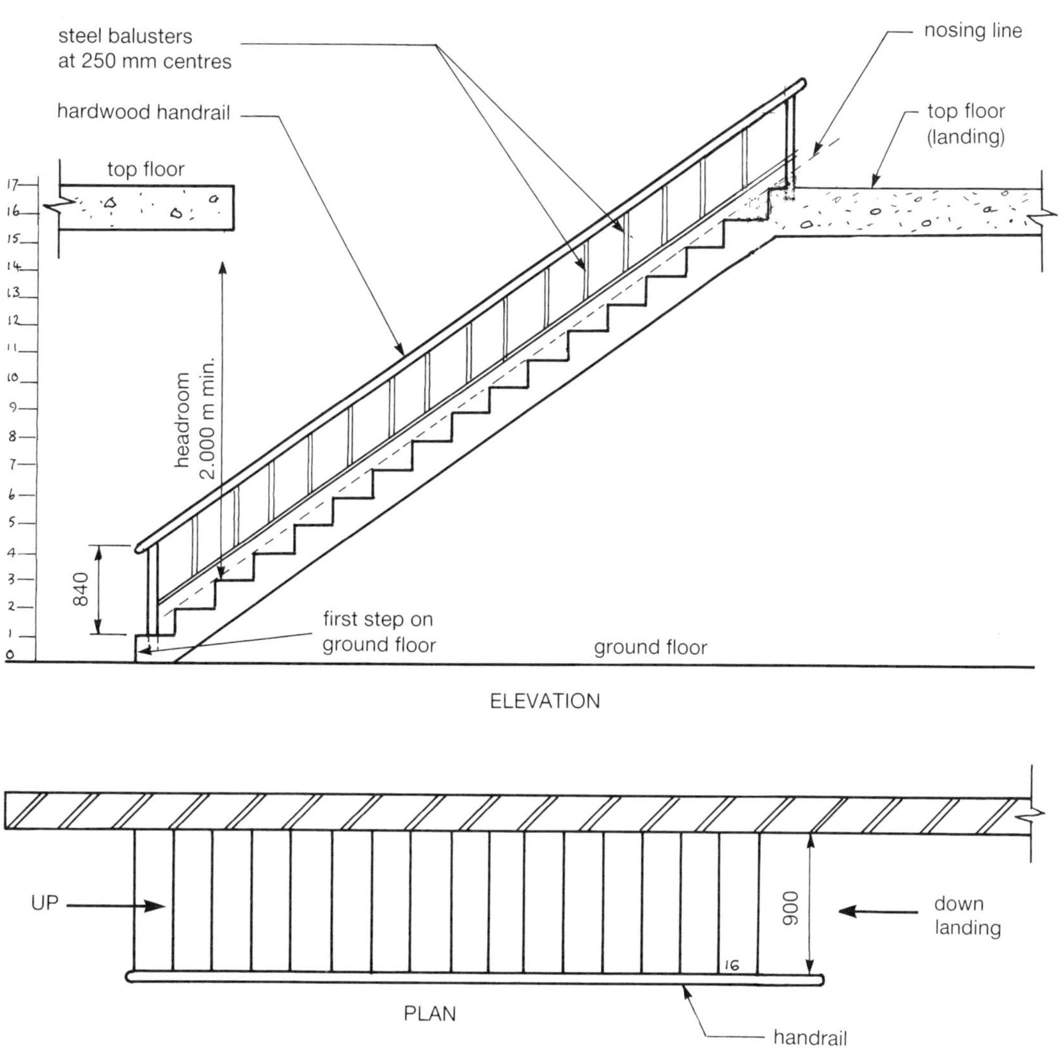
steel balusters
at 250 mm centres
hardwood handrail
top floor
nosing line
top floor
(landing)
headroom
2.000 m min.
840
first step on
ground floor
ground floor
ELEVATION
UP
900
down
landing
16
PLAN
handrail

Straight flight stair–details of balustrade

8. From the top of the step on which the newel post is located, measure out the height of the newel post such that when a 50 mm-thick handrail is drawn the total height from the step to the handrail is 840 mm. Note that the handrail should run parallel to the nosing line.
9. Draw the handrail and the balusters, spacing them at about 250 mm centres.
10. Erase all construction lines. Dimension and label all details as appropriate.

Drawing a stair with a landing and a turn

In most public and domestic buildings there are stairs that have quarter-turns, half-turns or open wells. The drawing procedure for these types of stairs is not different from the procedure for drawing straight-flight stairs. The major difference lies in the fact that these latter types involve a change of direction of either a 90° or 180° turn. They always have a landing at an intermediate position between the ground and top floors or between two upper floors. We shall draw one type in order to understand the procedures involved.

The dog-leg or half-turn stair

We earlier described the dog-leg stair as a stair having one flight rising to an intermediate half-space landing with the second flight running in the opposite direction to the first and parallel to it.

TECHNICAL DRAWING 3: BUILDING DRAWING

Suppose we have a space measuring 3.150 m long, 1.800 m wide and 3.000 m high, for purposes of constructing a dog-leg stair. We adopt the following steps:

1. Since we shall have two flights, we determine the number of steps in each flight. A freehand sketch of the plan view will be a useful guide (see right).

   The formula for calculating the tread and riser dimensions is $G + 2R = 550$ to 700 mm. Since we are given the height from floor to floor to be 3 m, we can make 17 steps with each rise measuring approximately 176.470 mm. This means we shall have 16 treads (goings), each of which should be about 250 mm. Providing for a landing of 900 mm, we have a total going of $(2.250 + 0.900)$ m $= 3.150$ m. Let us check to see if the rise and going we have chosen agree with our formula:

   $$\begin{aligned} \text{rise} &= 176.47 \text{ mm} \\ \text{going} &= 250 \text{ mm} \\ G + 2R &= 250 + 2(176.47) \text{ mm} \\ &= 602.94 \text{ mm}. \end{aligned}$$

   This is within the range 550 to 700 mm.

2. Having determined the dimensions of the rise, going and landing, we are now prepared to lay out the stair. First, choose a convenient scale for the drawing. If the stair is going to be a part of the cross-section of a building, the scale shall be the same as that for the entire section, say 1:50. If it is being drawn separately to emphasise the details, a scale of 1:25 may be appropriate. We shall draw separate stair details for purposes of illustration.

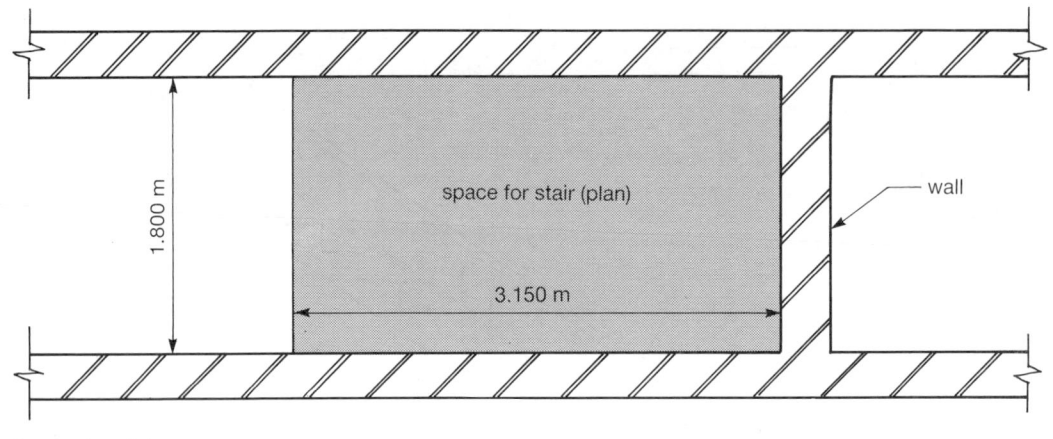

Space for stairs

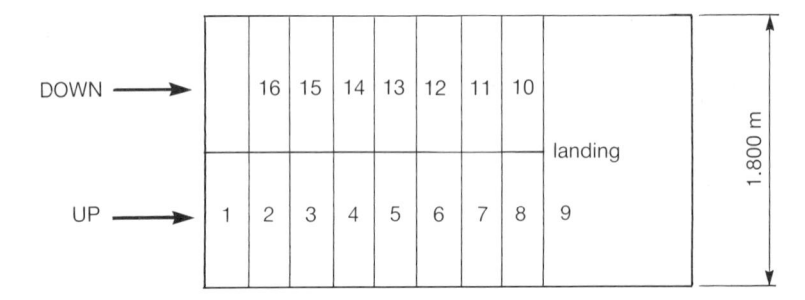

Determining the number of steps

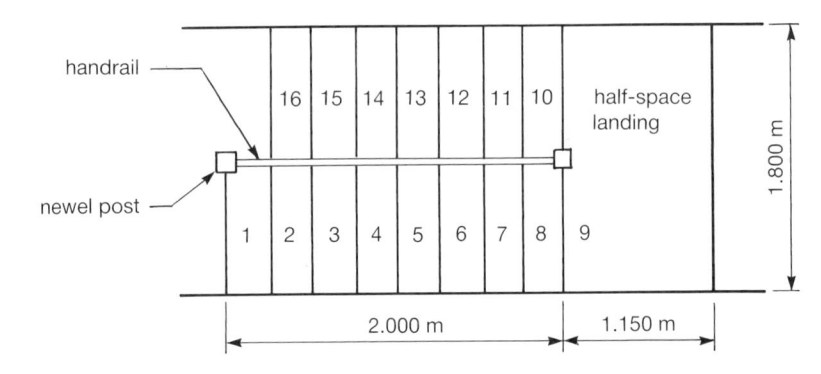

3. At the lower part of the drawing sheet, lay out and draw the plan of the stair. The elevation of the stair shall later be projected from this plan but sometimes, especially when drawing the sectional elevation of a building drawing, the plan of the stair may not only be unnecessary but impossible because of lack of space for it on the drawing sheet. To draw the plan:
   a) draw the plan area occupied by the stair to measure 3.150 m by 1.800 m,
   b) mark out the half-space landing to measure 1.150 m wide. Note that this width is the sum of one going (250 mm) and the landing (900 mm),
   c) divide the remaining 2 m space into eight equal parts of 250 mm each,
   d) locate the handrail 75 mm wide and the newel posts 100 mm square at the centre of the well,
   e) number the steps as shown on page 68.
4. From the plan project upwards to locate the elevation area of 3 m by 3.150 m. Project the treads also.
5. Divide the total rise of 3 m into 17 equal parts. This will produce the checkered area from which you should thicken the stair as shown.
6. Locate the newel posts at the beginning of the flight, the landing and the end of the flight on the top floor, as shown on the right.
7. Add the balustrade, that is the handrail and balusters, as with the straight-flight stair and as shown.
8. Erase all construction lines; label and dimension all details as appropriate.

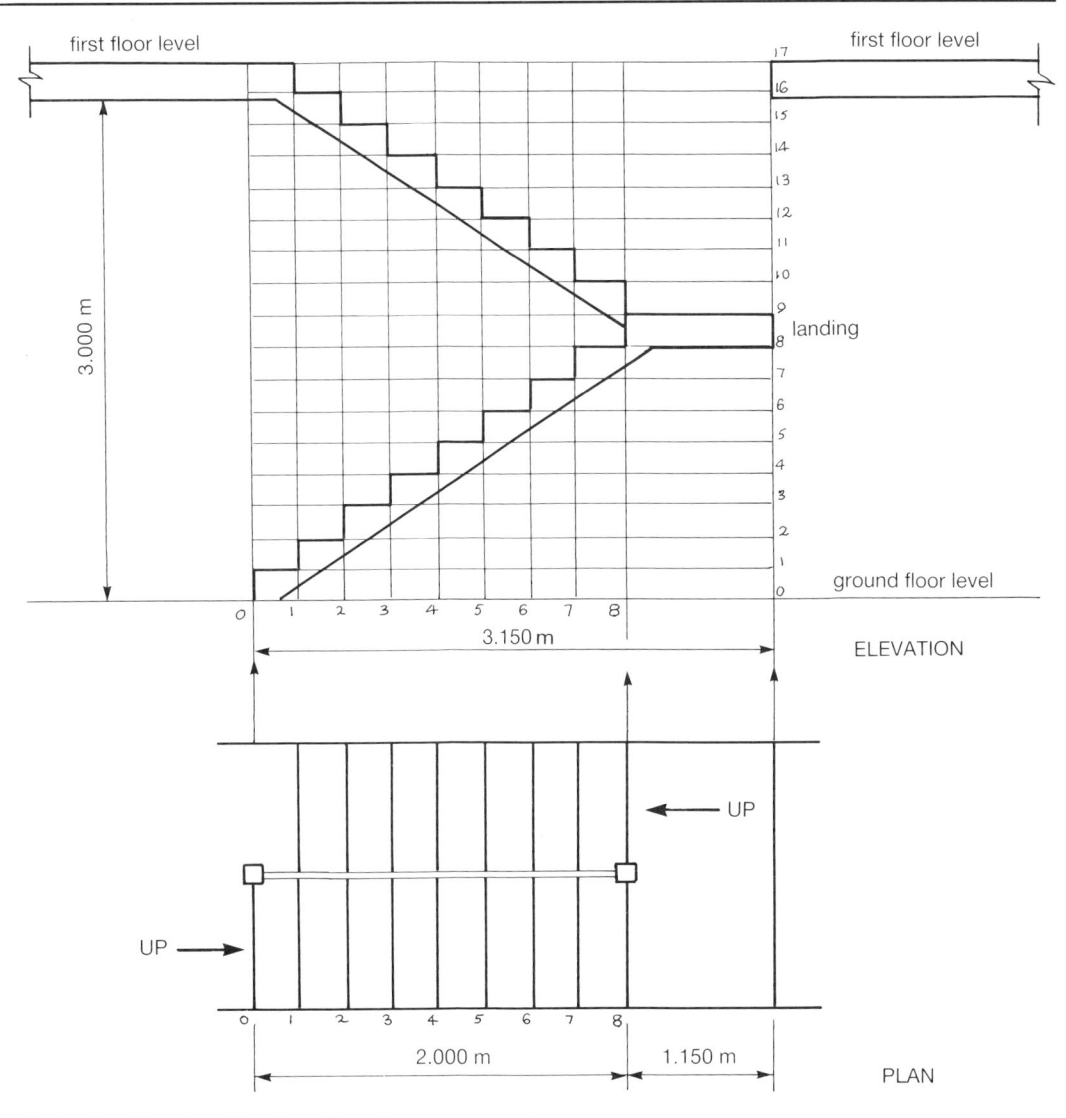

Projecting the elevation from the plan

TECHNICAL DRAWING 3: BUILDING DRAWING

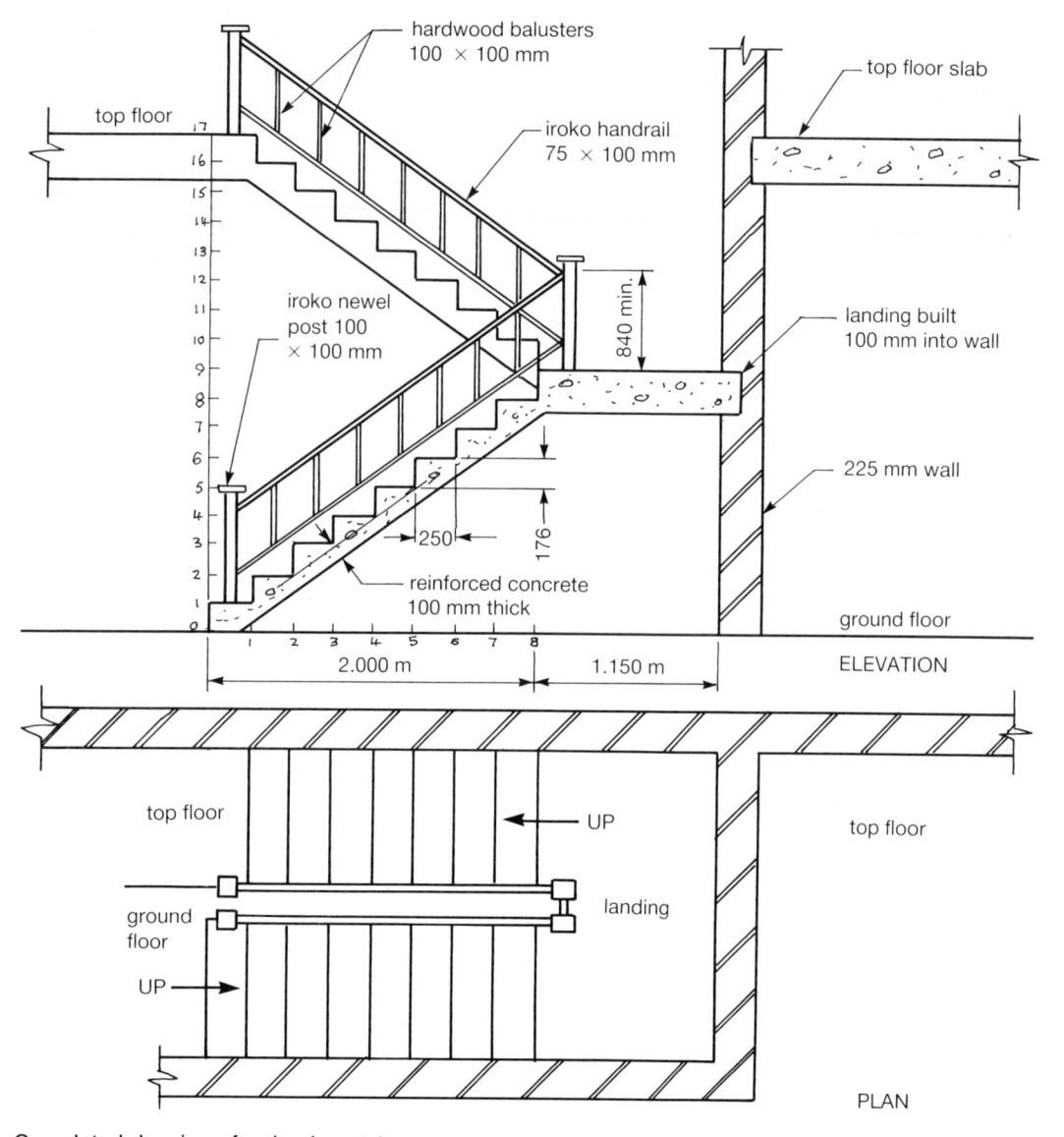

Completed drawing of a dog-leg stair

Exercises

1. Make neat sketches to distinguish a dog-leg stair from an open-well stair.
2. Draw a sectional detail of a straight-flight stair on a scale of 1:50, given that:

    rise = 185 mm,
    going = 225 mm,
    total going = 3.600 m,
    total rise = 3.150 m.

3. Draw the details of a dog-leg stair given the following specifications:
    stairwell is 3.150 m long, 1.800 m wide and 3 m high;
    landing = 900 mm;
    use design formula to obtain appropriate rise and going;
    scale 1:50.
4. Use neat sketches to define the following terms associated with stairs:
    a) baluster,
    b) balustrade,
    c) tread,
    d) riser.
5. Make a neat sketch of a step in a flight of stairs and in it distinguish between the following:
    a) rise and riser,
    b) tread and going.

10. Roofs

The roof of a house refers to the framework of timber, steel or concrete on which a covering of thatch, corrugated sheets, asphalt etc. is attached. The common components of a roof are shown on the right.

Components of a roof

1. *Common rafter*: A sloping timber extending from the eaves to the ridge of a roof.
2. *Eaves*: The overhanging or lowest part of a sloping roof.
3. *Gable end*: A gable end is formed when the wall is carried up to the underside of the roof.
4. *Hip*: The edge or angle formed when two roof surfaces meet to form an external angle which exceeds 180°.
5. *Hipped end*: The sloping triangular end of a hipped roof.
6. *Jack rafter*: A short rafter between the hip rafter and eaves or between the valley rafter and ridge.
7. *Overhang*: Extension of the roof covering between the outer face of the wall and the eaves line or verge.
8. *Pitch*: The ratio of the rise to the span:

$$\text{pitch} = \frac{\text{rise}}{\text{span}}.$$

9. *Purlins*: Horizontal timbers in a roof at right angles to rafters and carried on

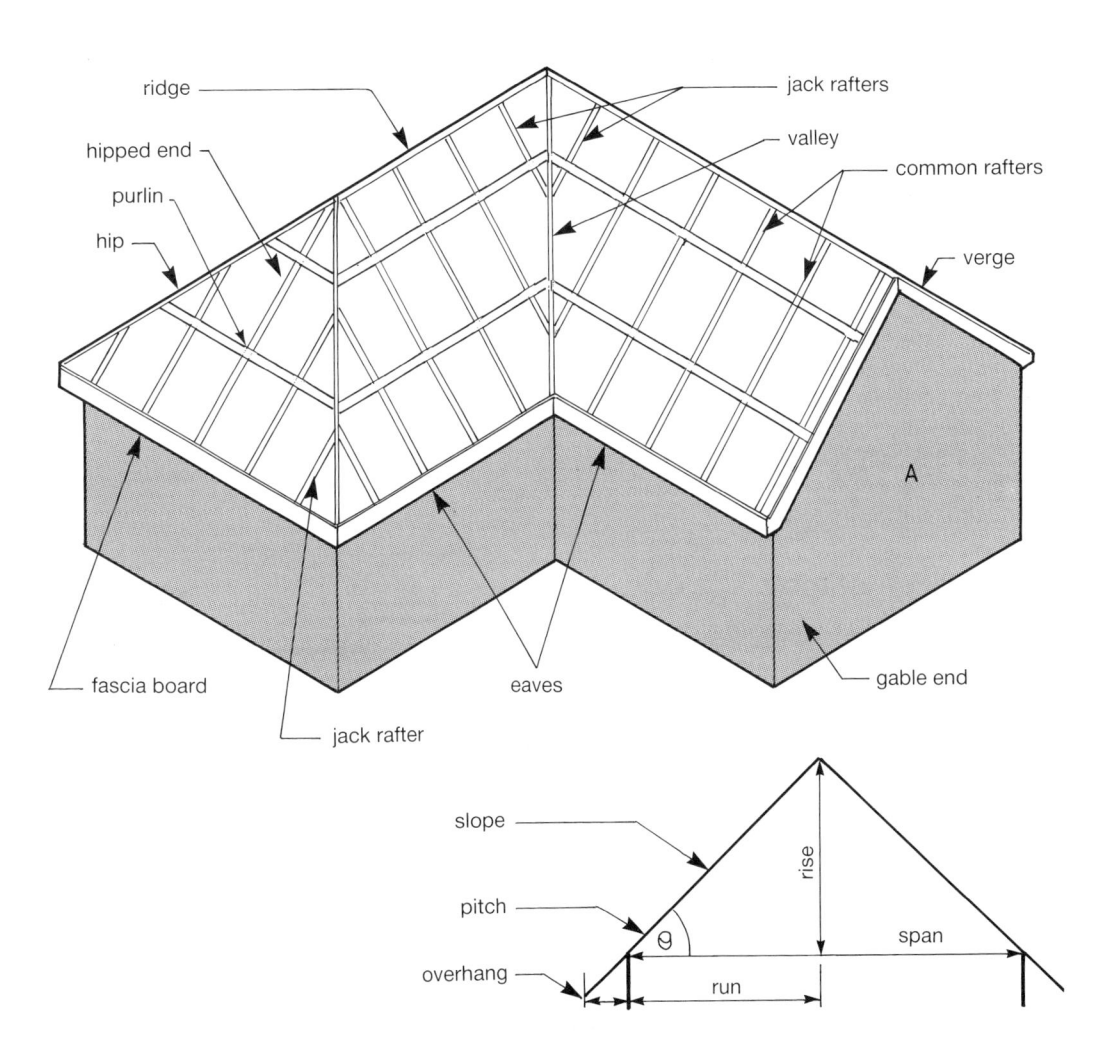

them to support a lightweight roof covering.

10. *Ridge*: The highest part or apex of a roof. The horizontal piece of timber forming the ridge and at which the common rafters meet is called the ridge piece.
11. *Rise*: The vertical distance from the wall plate to the ridge.
12. *Run*: One-half of the span. It is measured from the centre of the span to the inside of either wall.
13. *Slope*: This is usually given in degrees and refers to the inclination of the roof to the horizontal.
14. *Span*: The clear distance between walls in-to-in.
15. *Valley*: A valley; the opposite of a hip, is formed when two sloping roof surfaces meet to form an internal angle towards which water flows.
16. *Verge*: The edge of a sloping roof which overhangs a gable.

Classification of roofs

Roofs are classified as either flat or pitched. Flat roofs have their pitch not exceeding 5°. Pitched roofs have pitches over 5°.

Flat roofs

Flat roofs are usually constructed of either timber or concrete. The construction details of timber and concrete flat roofs are shown on the right. The essential components of a timber flat roof include 150 mm × 50 mm roof joists which are built into and bear on walls, 25 mm-thick boarding, firring pieces tapered to the required fall and a covering of asphalt or built-up felt.

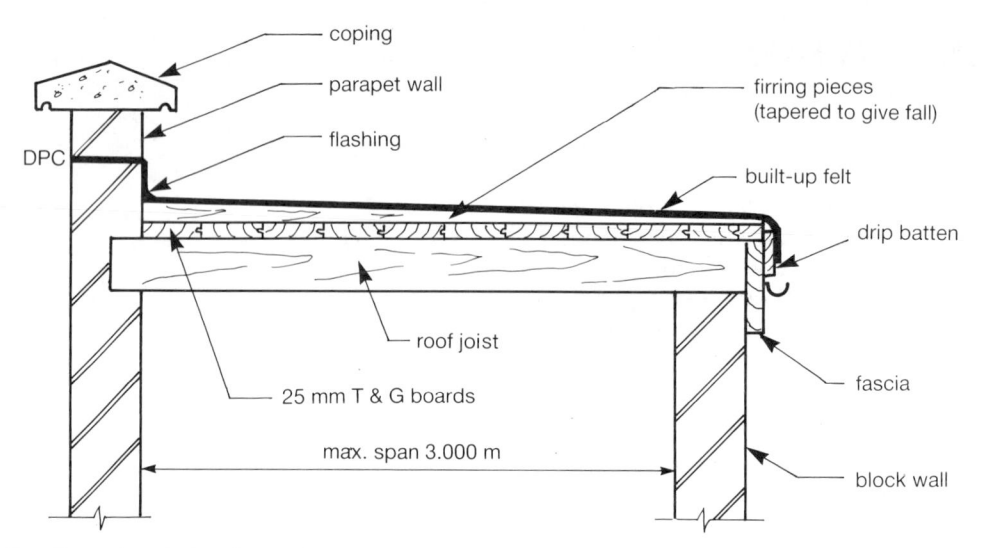

Section through a timber flat roof

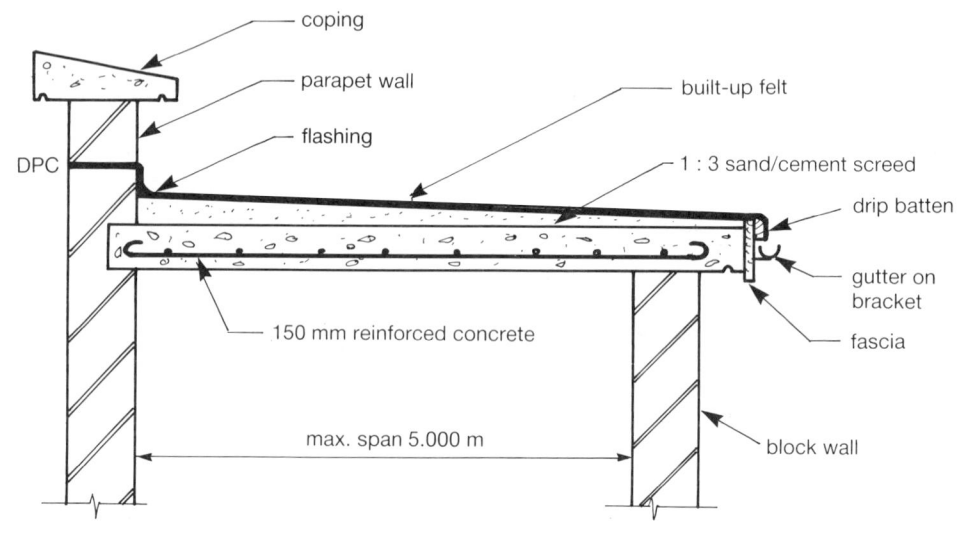

Section through a concrete flat roof

ROOFS

A parapet wall which extends beyond the roof improves the appearance of the roof, which otherwise would make the building appear incomplete. A 225 mm x 25 mm fascia, a continuous drip batten and a gutter all serve to ensure that water is thrown off the wall.

A concrete flat roof consists of a slab of reinforced concrete, 1:3 cement/sand screed to impart the required fall or slope, and a waterproof covering such as built-up felt. These features are usually best shown in the sectional elevation of a building.

In order that the roof may throw off water adequately the verge detail of a concrete roof is designed as below.

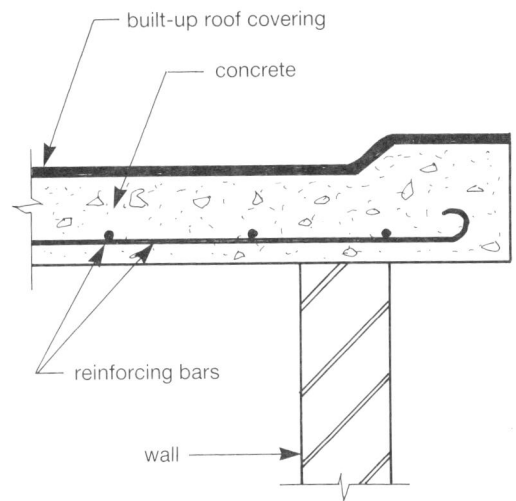

Verge details

Pitched roofs
Various styles of pitched roof shapes are used. Some of the common ones are shown on the right.

Monopitch roof

Gable roof

Hipped roof

Dutch gable roof

Saw tooth roof for factories

Folded plate roof for factories

73

TECHNICAL DRAWING 3: BUILDING DRAWING

Roof details in a building drawing

Apart from the floor plan, other views of a building show the roof. Taking a cross-section through a building shows details of the roof members and their relative positions on the roof as shown below.

Drawing a roof in elevation

The drawing on page 77 shows how the roof is drawn in the front elevation. Note the

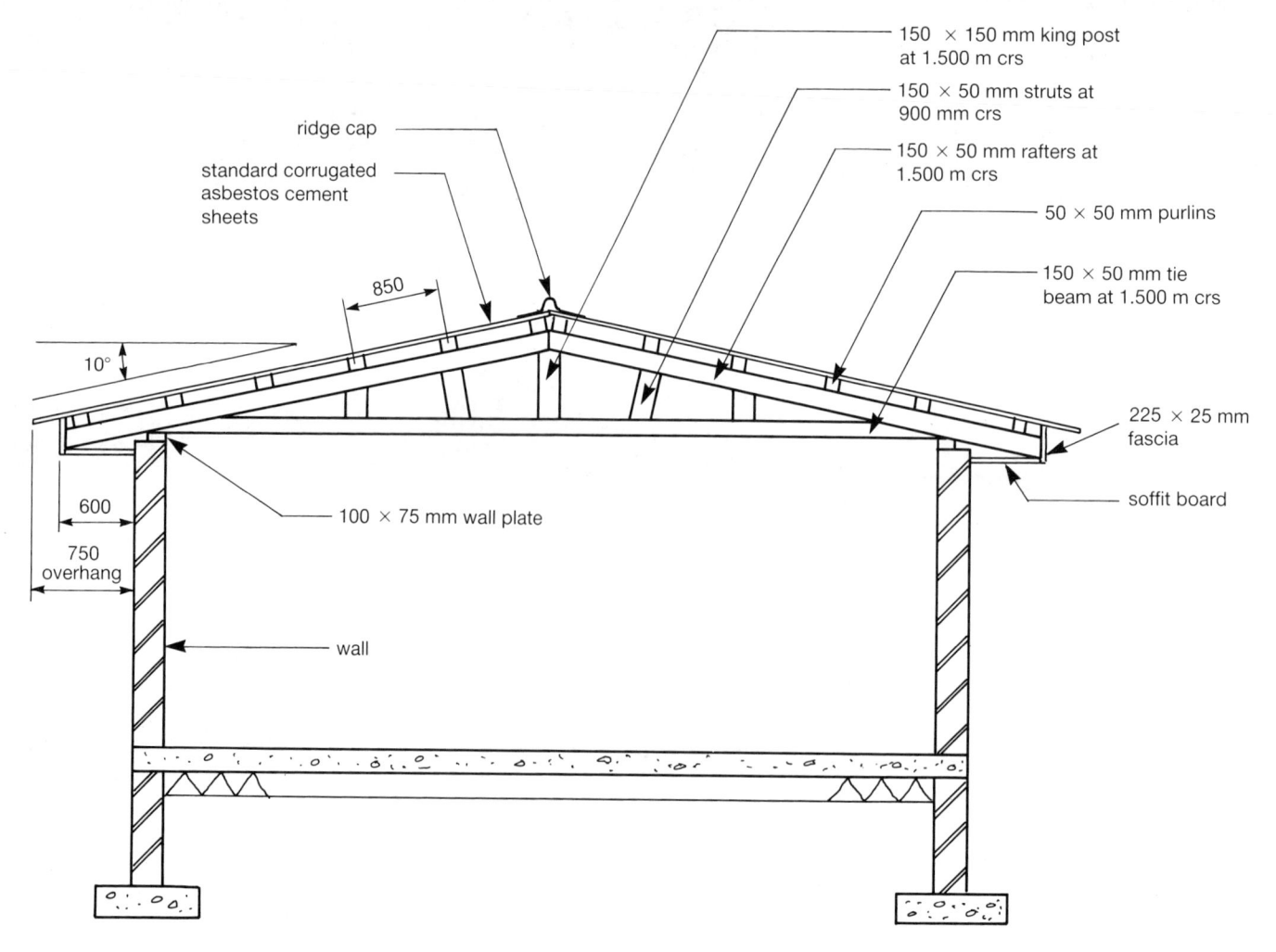

Section through a building showing roof details

74

ROOFS

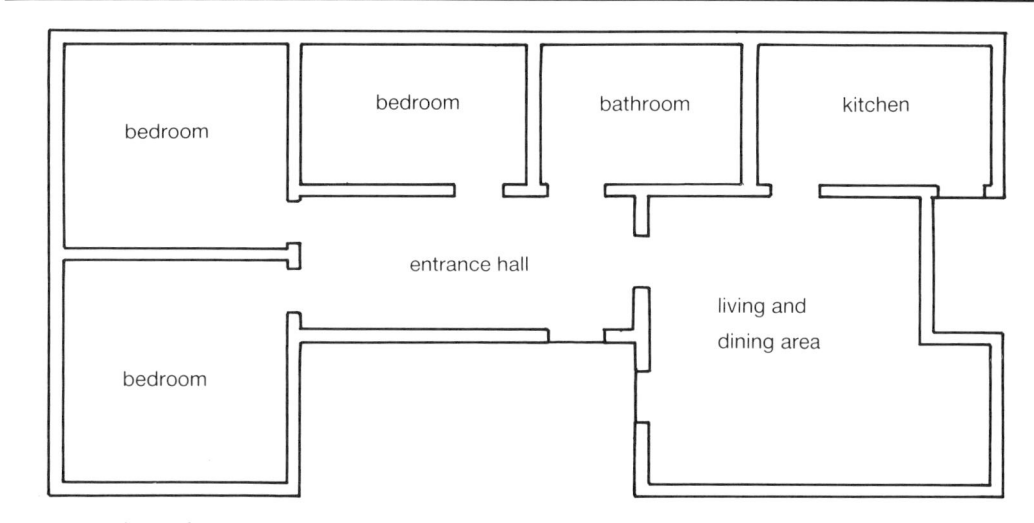

projection lines. It is not conventional to dimension the front and end elevations in a building drawing.

Roof plan

Sometimes it is required that the roof plan be shown. It is mostly the case whenever the building has a rambling shape. For simple gables or hips the roof plan may not be necessary. Let us take the floor plan on the left as an example of a building with a rambling shape. Even though we might use a gable roof for the floor plan, it may be necessary to draw this roof plan in order to guide

A rambling floor plan

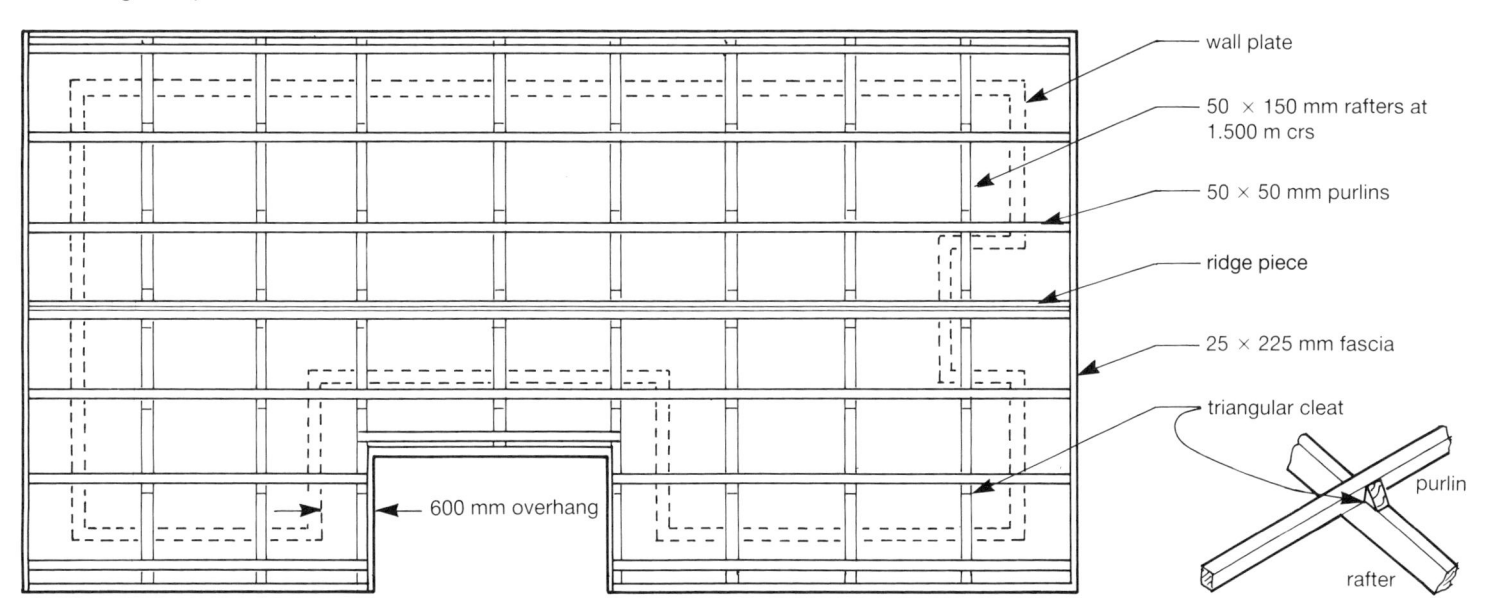

- wall plate
- 50 × 150 mm rafters at 1.500 m crs
- 50 × 50 mm purlins
- ridge piece
- 25 × 225 mm fascia
- triangular cleat
- purlin
- rafter

600 mm overhang

The roof plan for the floor plan shown above

75

TECHNICAL DRAWING 3: BUILDING DRAWING

FRONT ELEVATION   END ELEVATION

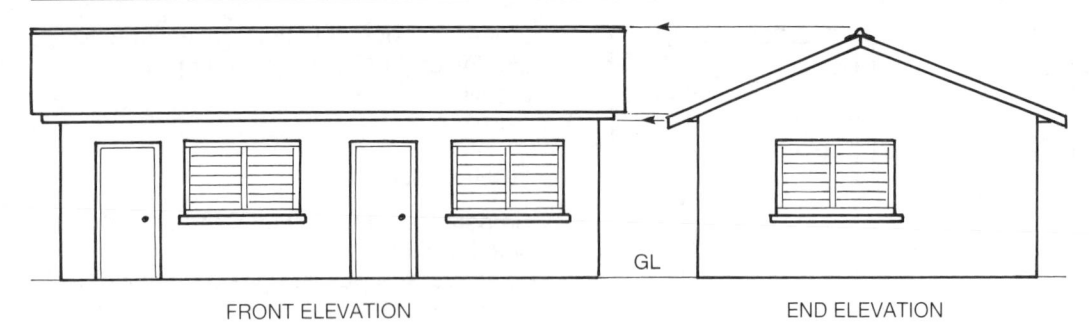

Projecting the front elevation from the end elevation

the builder. Such a roof plan will look as shown. Note that the roof plan may not necessarily take the shape of the floor plan but must be guided by the external walls.

Note that the walls are shown in dotted lines. The purlins are supported on the rafters by triangular cleats nailed to the rafters, the shape and size of which are shown on page 76.

Exercises

1. Sketch a simple roof and label the following parts:
   a) ridge piece,
   b) common rafter,
   c) eaves,
   d) purlins,
   e) gable end.
2. a) How are roofs classified?
   b) Which of the following is not a pitched roof angle:
   i) 30°, ii) 12°, iii) 3½°, iv) 7½°, v) 6°?
3. Sketch a timber flat roof showing the location of:
   a) the roof joist,
   b) the firring pieces.
4. Identify, using sketches, any five pitched roof styles.
5. Which of the following drawings will *most likely* reveal the main components of the roof of a building:
   a) floor plan,
   b) front elevation,
   c) end elevation,
   d) sectional end elevation,
   e) roof plan?
6. What determines the shape of the roof plan of a building?

11. Electrical wiring plans

Electricity in buildings

Windows and transparent roofing sheets allow light into buildings only during the daytime. However, people also need light in houses at night. While hurricane lamps, candles and battery operated lamps could be used as sources of light at night, they are usually not as durable and versatile as sources using electrical energy from the mains.

Electrical energy is needed in homes for lighting and for operating radios, televisions, video recorders, refrigerators, electric cookers, washing machines and other household appliances.

Types of lamps used in buildings

Two main categories of lamps are used in lighting buildings: incandescent and fluorescent lamps.

Incandescent lamps

An incandescent lamp consists of a glass bulb which holds a metallic filament often made of tungsten. The filament emits light when an electric current is passed through it. Such lights are made for a variety of uses, including the lighting of rooms. Incandescent lamps can be fixed on ceilings or walls. They can also be fixed in rooms or used outside the rooms as security lamps.

Fluorescent lamps

Fluorescent lamps are widely used in all types of buildings. They are designed to run on alternating current (AC) supplies but can be adapted to direct current (DC) supplies if desired. Fluorescent tubes can be fixed on ceilings or on walls, and both indoors and outside.

Components of an electrical wiring plan

The major components of an electrical wiring plan include the following:
1. switches, showing their locations, using appropriate symbols,
2. wall socket outlets and their location properly represented with symbols,
3. the main switch,
4. meter box(es).
5. lamps showing types and locations.

As the name suggests, the electrical plan is a drawing showing the plan of the building together with the location of switches, meter boxes, lamps, etc. It is important to plan the electrical wiring details in order to ensure safe and efficient operation of the electrical system and reduce the rate of maintenance in the system.

Electrical wiring symbols

Some of the symbols used in electrical wiring drawings are shown on page 79. These symbols are based on BS PD 7303 (1981).

Locating electrical components on a wiring plan

Service entrance

The service entrance or service line usually starts from a power supply line (e.g. a power line or a generator). This line should not run for more than 5–7 m inside the house before it reaches the main disconnect or isolator switch. The closer the main switch is to the meter, the better. Also the service entrance should be located close to where the greatest, amount of electricity is consumed to reduce the amount and cost of wiring, for example, the kitchen. The meter and distribution board should also be located close to areas of greatest usage of electricity. A meter located outside the house makes for easier reading.

Switches

The number of switches and their location is affected by the number of lighting fixtures, socket outlets, and other equipment used. In locating switches,

77

consider the most logical location for each switch and the traffic pattern in the house. The electrical wiring plan should show the type of switch using the appropriate symbol. Several types of switches are used in houses and some are preferred to others for their unique qualities. Switches used in electrical wiring include the following:
1. toggle switch,
2. rocker switch,
3. push-button switch,
4. dimmer switch,
5. delayed-action switch.

On the plan, switches are shown connected with dotted lines to the fixtures, appliances or outlets which they operate. These dotted lines do not represent wiring lines but merely indicate the fixture operated by a given switch. These lines should be drawn with irregular (French) curves rather than as straight lines using a straight edge, or freehand. Freehand lines are not attractive and straight lines tend to become confused with other lines.

LIGHTING SYMBOLS

×	lighting point or lamp
× 3 × 60 W	three lamps at 60 W
⋈	wall-mounted lamp or lighting point
×	lighting point with built-in switch
⊢—⊣	fluorescent lamp
⊢—⊣ 2 × 65 W	group of two fluorescent lamps at 65 W each

LIGHTING SWITCHES

♂	one-way switch
♂↓	chord operated one-way switch
♂♂	two-way switch
♂₃	three single pole switches together

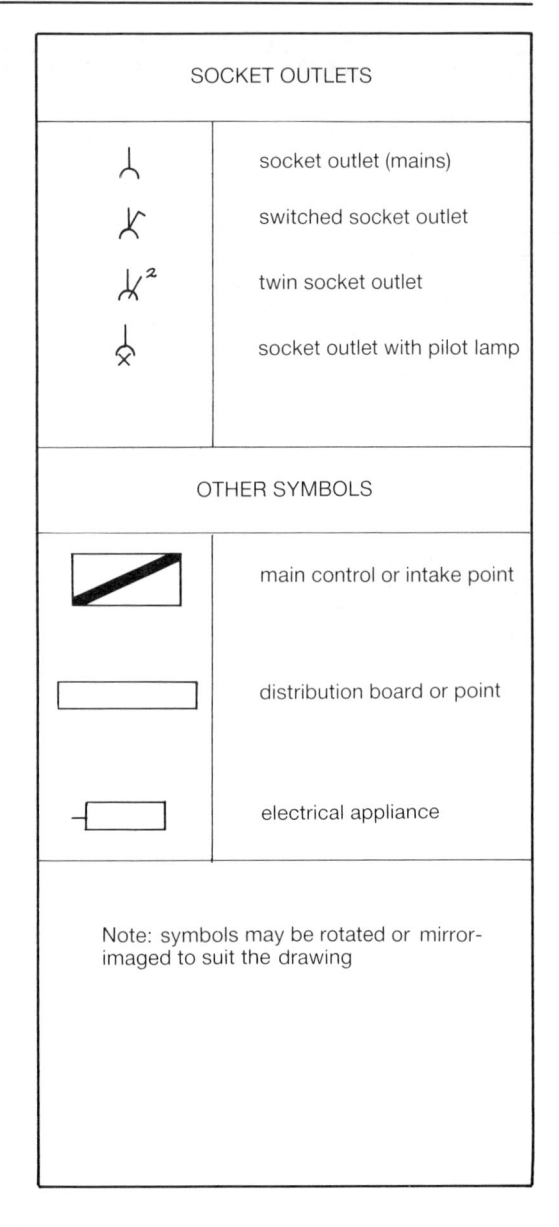

SOCKET OUTLETS

⊥	socket outlet (mains)
⊬	switched socket outlet
⊬²	twin socket outlet
⊥	socket outlet with pilot lamp

OTHER SYMBOLS

▨	main control or intake point
▭	distribution board or point
⊣▭	electrical appliance

Note: symbols may be rotated or mirror-imaged to suit the drawing

Socket outlets

Socket outlets used in houses are mainly of two types–switched outlets and unswitched (live) outlets. Outlets without switches remain on all the time and hence are said to be 'live'. Most outlets used are of the 220–240 V switched type. Twin sockets (two together) are also available.

In locating socket outlets it is recommended that they be spaced about 2–3 m apart in all rooms, and that consideration be given to the arrangement of appliances and furniture. All socket outlets must be earthed for reasons of safety.

Lighting required in houses

There is no specific amount of light in a room which can be said to be suitable to all occupants. The important factor influencing the amount of light needed in a room is the kind of activity generally carried out there.

In locating a lighting fixture in a room consideration must be given to the following:

1. Does the bulb or tube require some shielding to reduce glare? When shading is necessary, diffusing bowls and shades are used to reduce glare.
2. Is the fixture to be permanently mounted? Usually ceilings and walls provide good mounting locations for permanently mounted light fixtures. Some types of lamp (table and pedestal) are simply plugged into socket outlets.

The following are some of the trends in locating lighting fixtures in houses:

1. fewer ceiling-mounted fixtures but more lamps plugged into outlets,
2. ceiling fixtures in the dining room that are centred over the dining table,
3. use of fluorescent tubes above a suspended ceiling in the kitchen and bathrooms. This technique provides both more light and an attractive finish,
4. use of recessed lighting fixtures in hallways and special emphasis areas.

Lighting fixtures must be represented with appropriate symbols drawn in the desired location of the fixture. It is also recommended that a lighting fixture schedule be included with the electrical plan. The schedule should show:

1. type of lighting fixture desired,
2. number of each type of lighting fixture needed,
3. mounting height and location,
4. the wattage of the fixture,
5. any other important information that will make for safe and efficient utilisation of the fixture.

Steps in drawing the wiring plan

1. Draw the floor plan showing all external and internal walls and major appliances.
2. Locate the meter and distribution board, showing their voltage and amperage ratings.
3. Locate all socket outlets using appropriate symbols.
4. Locate all ceiling and wall lighting outlets. Be sure to use proper symbols.
5. Show all special outlets and fixtures like telephones, door bells etc.
6. Locate all switches and link them up (using dotted lines drawn with irregular curves) to the lighting fixtures which they operate.
7. Add the lighting fixture schedule (specification) and symbol legend.
8. Put in every piece of information desired in the title block section, e.g. title of drawing, scale, person who made the drawing, sheet number etc.
9. Check the electrical drawing in terms of appropriateness of its design, as well as for the accuracy and completeness of the information supplied.

An electrical wiring plan in which the above steps have been applied is shown on page 81.

Exercises

1. Identify six major components of an electrical wiring plan.
2. Two important factors must be considered when locating a service entrance. What are they?
3. Name three types of switches that may be used in the home.
4. What are the correct symbols for the following:
   a) fluorescent tube,
   b) single-pole switch,
   c) single socket outlet,
   d) electrical appliance?
5. Draw a two-bedroom house showing the fixtures, outlets and appliances in:
   a) kitchen,
   b) bathroom,
   c) living/dining room,
   d) bedrooms,
   e) hallways and verandahs.

TECHNICAL DRAWING 3: BUILDING DRAWING

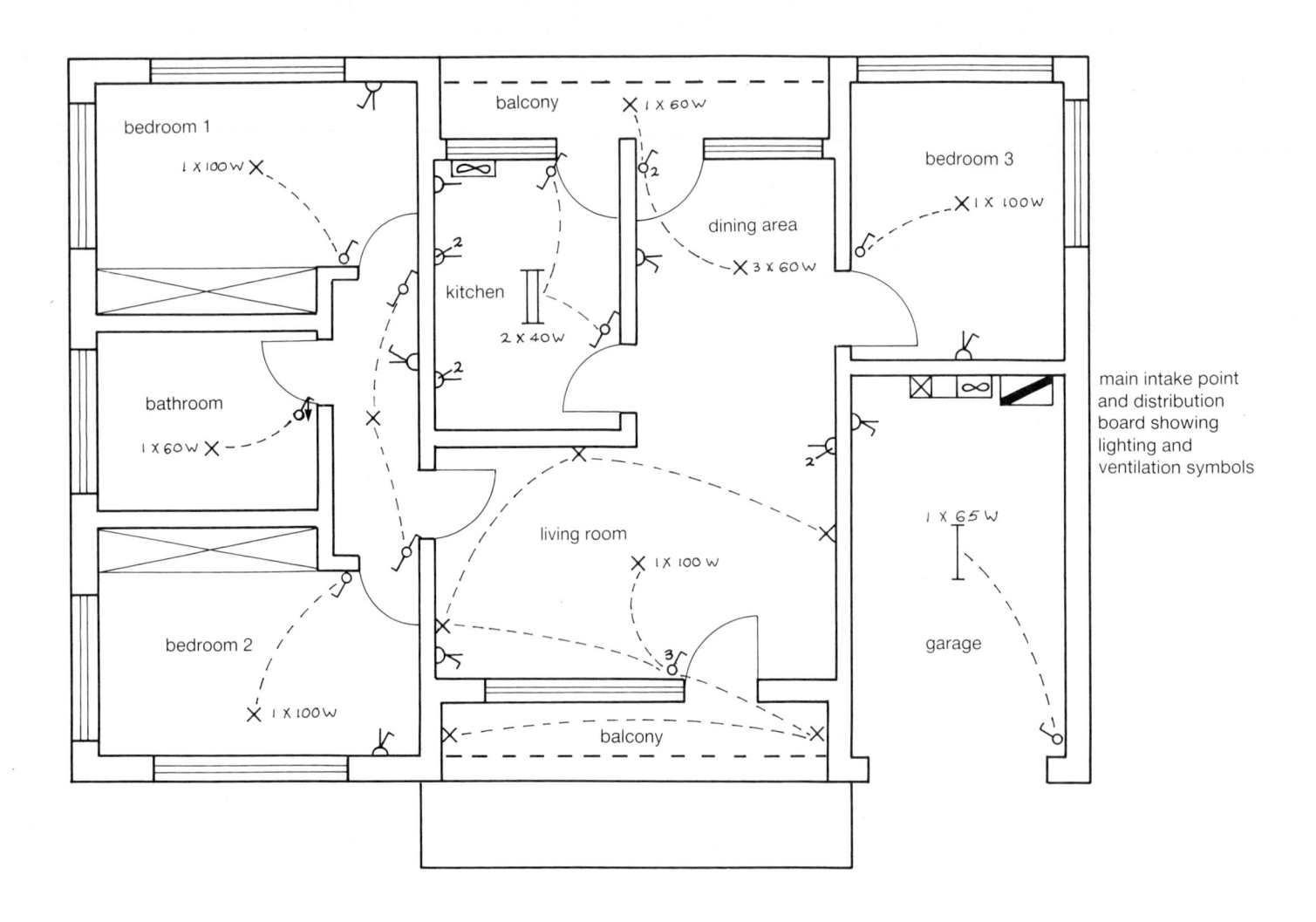

An electrical wiring plan

12. Drainage and sanitation systems

The design and construction of domestic and industrial buildings require adequate provision for a plumbing system. Plumbing work in buildings is very important, just as other services like the electricity and gas supplies. In buildings a plumbing system is concerned with the provision of an adequate supply of good quality water for drinking and other purposes and at desired locations.

A plumbing system is also needed for the removal of waste through sanitary appliances into a sewer or a private septic tank. Generally, the design and installation of a plumbing system involves three essential components:
1. supply of cold and hot water,
2. removal of water and waste,
3. choice of fixtures that facilitate the use of water.

Water supply systems

Water supplied to domestic and industrial buildings is usually generated by the city water works and supplied to the public through the city main. From the city main, individual buildings draw their water supply using communication pipes or the building main, laid at a convenient position and terminated with a stop valve. From the stop valve, service pipes are connected to the building or to the necessary draw-off points. In most cases, the service pipes also feed into a cistern, positioned on an elevated height near the building or on concrete roofs over kitchens.

Storing water in a cistern is done to prevent supply interruption that may be caused by maintenance works or repairs of burst pipes along the line. Cold water supply normally runs at high pressure. Therefore appropriate and sound fittings to withstand water pressure must be used.

Designing a plumbing system

In architectural drawing, it is important to incorporate detailed drawings of the plumbing works for a particular building. These are usually represented with appropriate symbols and conventions on the building plan (see page 83).

A well-designed plumbing system for a given building must consider the following:
1. the position and location of the water main in relation to the building site,
2. the most economical and efficient way of connecting the building to the main,
3. the sizes of pipes required at a particular connection,
4. the location of necessary sanitary fixtures in the building relative to the pipe run,
5. the pressure under which water is supplied,
6. local authority bye-laws on water services,
7. the code of practice and regulations regarding water services.

Designing and drawing a cold water supply system

In designing and drawing a cold water supply system, the following procedures are recommended:
1. Design the floor plan as required to meet the needs of the occupants of the proposed building.
2. Locate on the floor plan the water service areas such as bath, toilet, kitchen, dining area, laundry, shower and other draw-off points, as required.
3. Consider the direction for possible connection of the building to the public mains. Try to minimise the pipe run as much as possible. Also try to avoid running pipes under rooms of the building. Where pipes must run under rooms, metal pipes should be used.
4. Use appropriate conventions for all connections and fittings.
5. Tap and connect from the city main supply to the communication pipe or building main up to the site boundary or just beyond it. Finish off the building main with a stop valve. Cover the stop valve with a chamber. Usually 18–25 mm-diameter pipes are used to tap

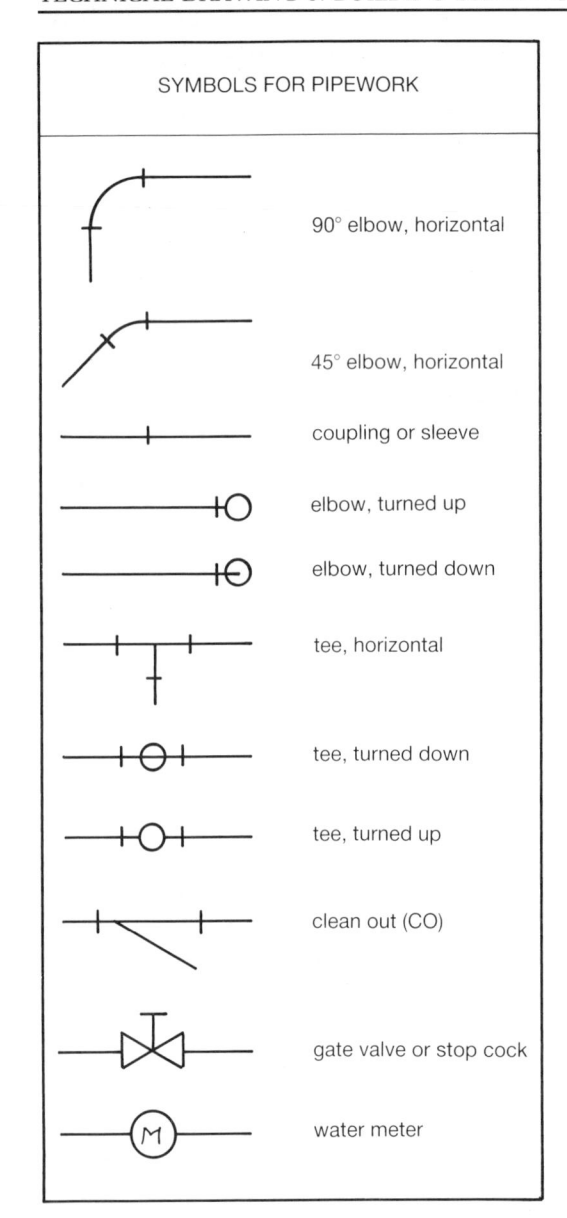

TECHNICAL DRAWING 3: BUILDING DRAWING

SYMBOLS FOR PIPEWORK

- 90° elbow, horizontal
- 45° elbow, horizontal
- coupling or sleeve
- elbow, turned up
- elbow, turned down
- tee, horizontal
- tee, turned down
- tee, turned up
- clean out (CO)
- gate valve or stop cock
- water meter

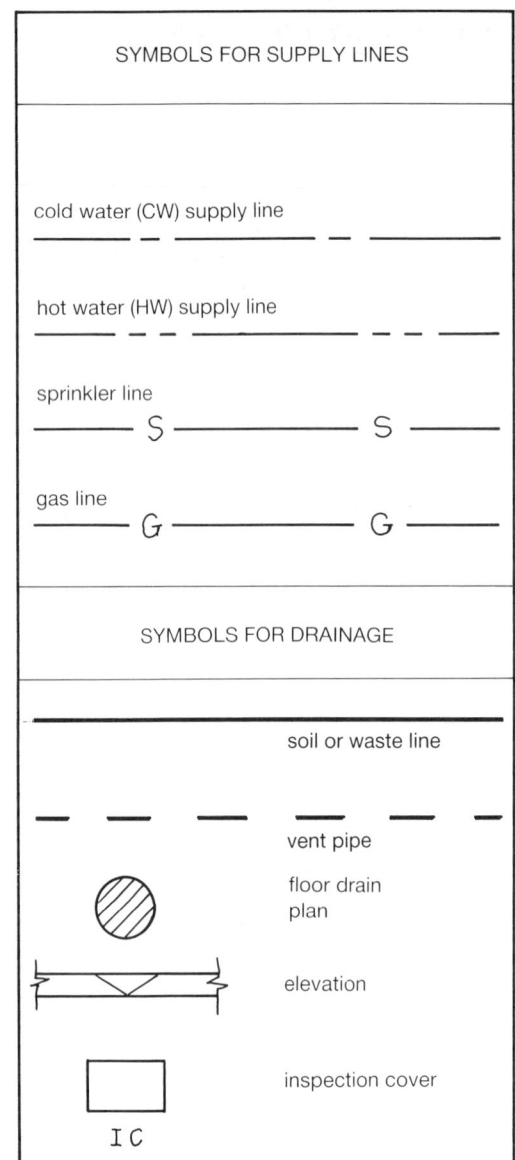

SYMBOLS FOR SUPPLY LINES

- cold water (CW) supply line
- hot water (HW) supply line
- sprinkler line — S — S —
- gas line — G — G —

SYMBOLS FOR DRAINAGE

- soil or waste line
- vent pipe
- floor drain plan
- elevation
- inspection cover (IC)

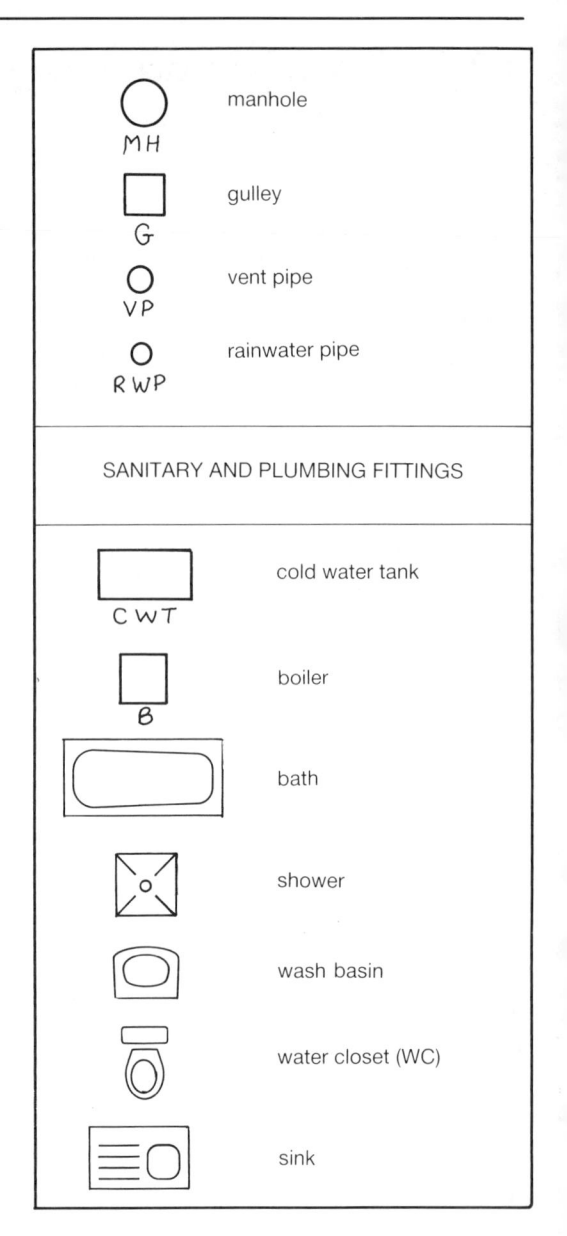

- manhole (MH)
- gulley (G)
- vent pipe (VP)
- rainwater pipe (RWP)

SANITARY AND PLUMBING FITTINGS

- cold water tank (CWT)
- boiler (B)
- bath
- shower
- wash basin
- water closet (WC)
- sink

water from the public mains to the building.
6. Connect an 18 mm-diameter pipe from the stop valve to the building. Then link the respective draw-off points with 13 mm-diameter pipes. At each draw-off point, use an appropriate stop valve to check the flow of water at that point without shutting off the water supply to other draw-off points.
7. If required, continue the service pipe to feed into a cistern for storage purposes. The cistern could be supported on a concrete flat roof, on pillars or on galvanised metal stanchion posts. Where a cistern is not necessary, continue the service pipe to the kitchen, toilet, bath, and other areas as required and connect water to these locations. The plan drawing on page 85 is of a cold water supply system.

Hot water supply systems

Occupants of a building usually need hot water in addition to cold water. Hot water production and distribution is not undertaken by the local water board or authority. Rather, individual buildings have independent methods of obtaining hot water at the desired temperature and quantity.

Hot water for domestic use is produced in two ways: central hot water supply and local or instantaneous hot water supply. In central hot water supply, cold water is fed into a boiler, where it is heated and then stored in a hot water storage tank for distribution to draw-off points called taps. In a localised hot water supply system, a heat source is positioned adjacent to the fitting where hot water is to be drawn. The water is heated and stored within that vicinity for use. The instantaneous hot water supply system has the major advantage of reducing pipe run. For effective design of a hot water system the following points should be considered:
1. sources of heat, which include solid fuel, gas and electricity,
2. the number of draw-off points for hot water,
3. the danger of fire outbreak,
4. insulation of storage tanks and pipe lines,
5. the height for locating the hot water storage tank,
6. the strength of fittings to withstand heat and pressure.

Drawing a hot water system
The following procedures are recommended for drawing a hot water system (see page 86):
1. Locate the source of heat, hot water storage tanks and the draw-off points.
2. Draw the hot water pipe line using the appropriate conventions. The hot water pipe line should be parallel to the cold water pipe line and about 150 mm apart; this is to avoid pipe congestion in the building.
3. Locate sanitary appliances near to the heat source and close to each other; short pipes minimise heat loss. The drawing on page 86 includes the layout of a hot water supply system.

Design and layout of a drainage system

Waste generated by household activities needs to be disposed of. Such waste includes human waste, in the form of excrement and foul water. Waste is removed from sanitary appliances using appropriate pipes or drains which lead into a sewer or septic tank and finally conveyed to a disposal unit. Pipes for removal of waste should be separated and isolated from pipes for the fresh water supply.

The flow of sewage in pipes is due to gravitation and not due to pressure. As a result every drain should have a sufficient fall to enable its content to flow easily and also be self-cleansing. A good knowledge of the theory and working of a drainage system is very important for determining:
1. the location and position of the public sewer in relation to the building site,
2. the possibility of connecting a private sewer to the public sewer without undue obstructions,
3. the local authority bye-laws affecting drainage systems,
4. the number and location of sanitary fixtures in the building,
5. the type and choice of pipes for the drainage system,
6. the suitability of the soil to be used for a disposal field.

Layout of a drainage system
The following procedures are helpful for drawing the layout of a drainage system (see page 87):
1. Draw to scale the floor plans showing exterior and interior walls, doors,

TECHNICAL DRAWING 3: BUILDING DRAWING

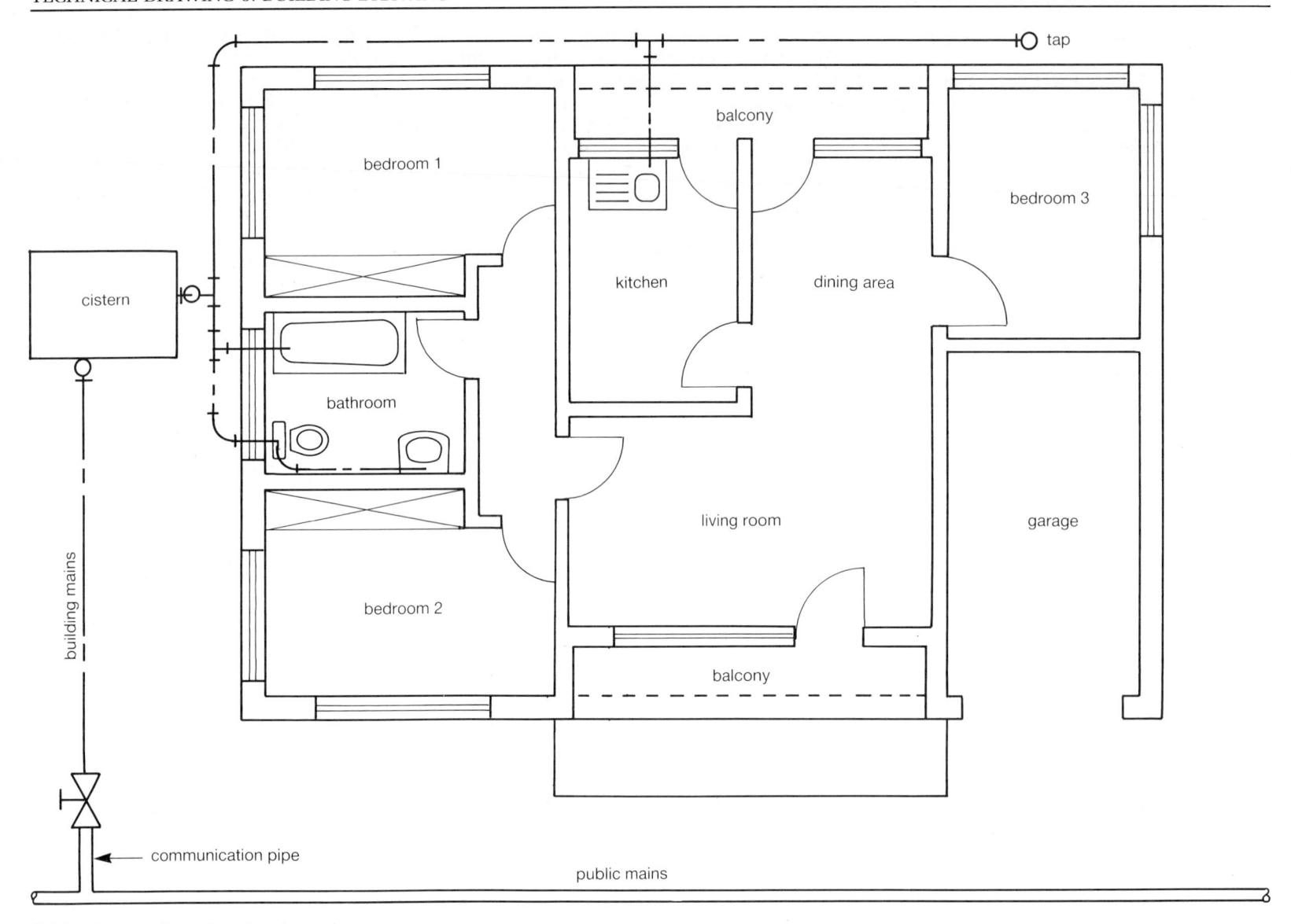

Cold water supply system in a bungalow

84

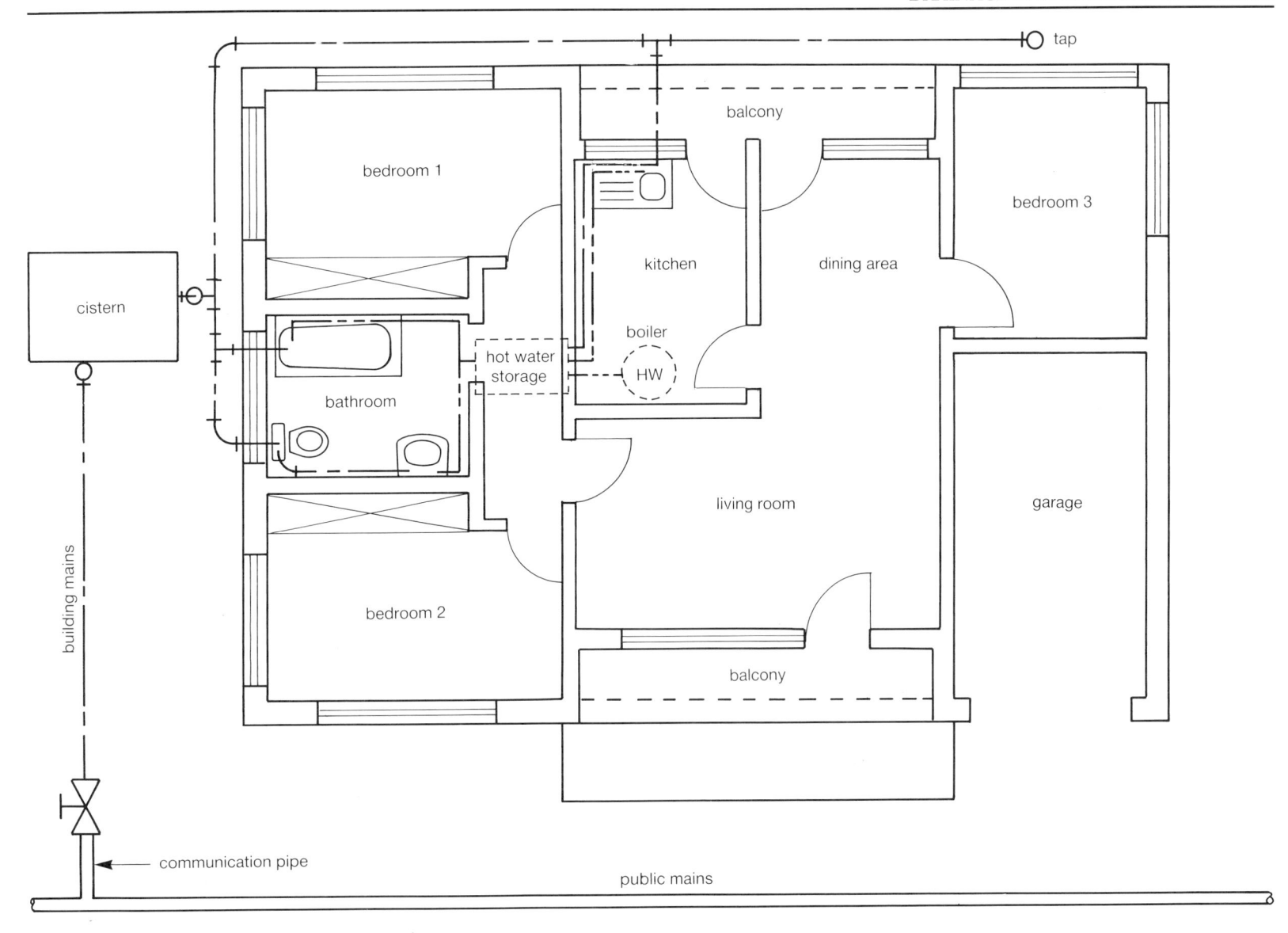

DRAINAGE AND SANITATION SYSTEMS

Cold and hot water supply systems in a bungalow

85

TECHNICAL DRAWING 3: BUILDING DRAWING

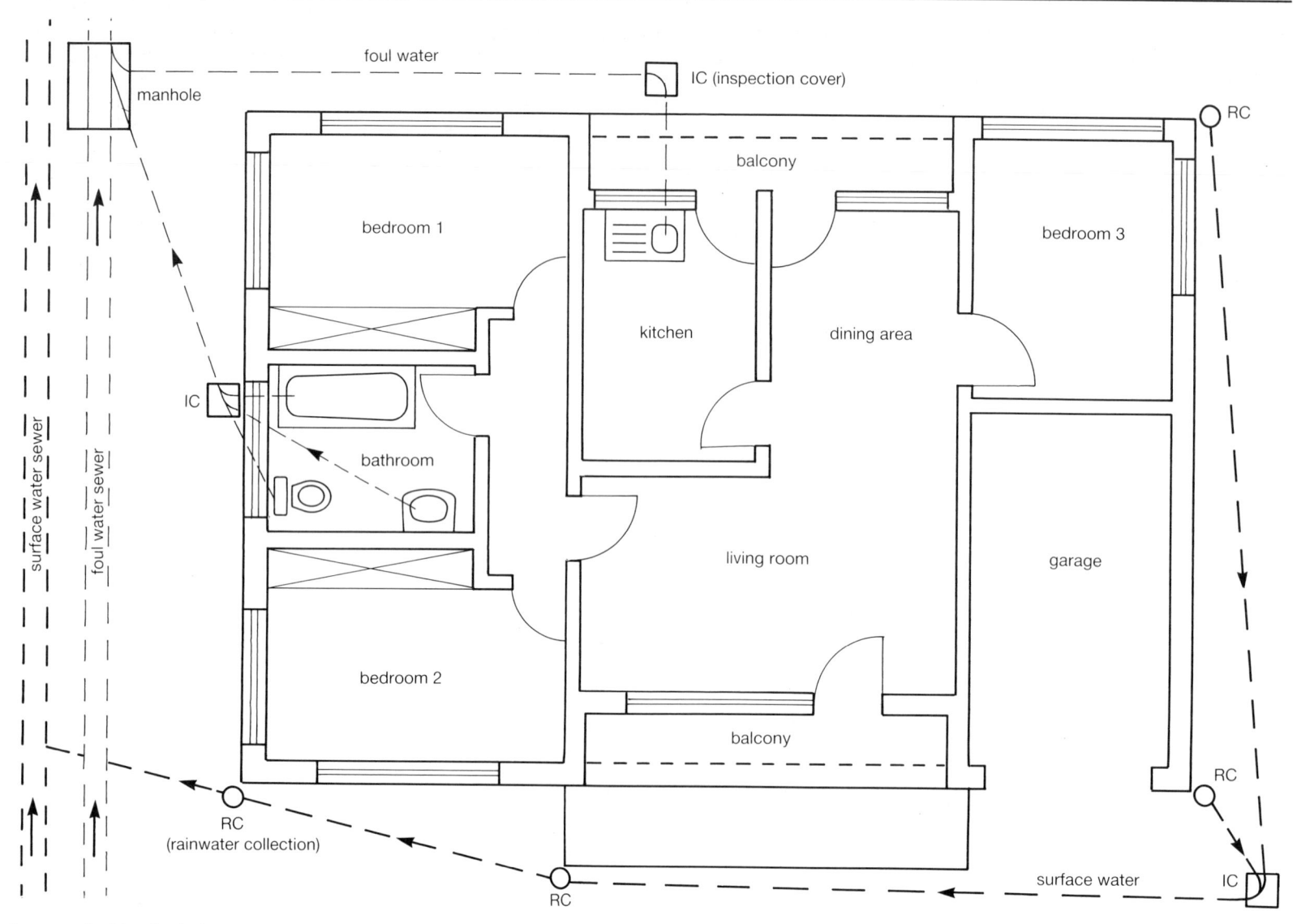

Layout of a simple drainage system in a bungalow

DRAINAGE AND SANITATION SYSTEMS

windows and other features in relation to the plumbing work.
2. Locate the positions and draw the outlines of all the sanitary fixtures to be drained. Use appropriate conventions and symbols to represent the fixtures and materials in the drainage system.
3. Determine the location and draw the building drain, soil stack and vent pipes. Be sure to draw all drains to desired falls (gradient).
4. Connect all floor drains (e.g. ground floor, first floor, etc.) to the building drain. Show all the necessary fittings. Also, specify the pipe diameter and type. Junctions to drains should not be at 90°.
5. Connect the building drain to the private sewer or to a public sewer, where it is available. At every change in direction, the bend should be as slow as possible, that is, not sharp.
6. Provide a legend for all symbols and conventions used for fittings and other materials in the drainage system.

Disposal of waste from buildings

Waste from sanitary appliances and drains in a building is usually deposited into a disposal unit. In domestic buildings a soakaway pit may be used. Liquid effluent from the septic tank drains through the soakaway pit into the adjoining soil.

To guide the construction of a disposal unit, septic tanks and soakaway pits are usually drawn to show the sections of these components and the materials used, including their manner of fixing or jointing.

Drawing a septic tank and soakaway pit
To draw the sectional details of a septic tank and soakaway pit:
1. Draw the water closet (WC) showing its connection through the wall, to an inspection chamber, located immediately outside an external wall behind the toilet. For this purpose, use a WC bend having a radius of 100 mm and which is connected to a 100 mm-diameter pipe. Be sure that no pipe joint terminates inside the wall (see page 89).
2. Draw a section of the inspection chamber, showing the direction of flow of the main channel. It is normal to use half-section channel pipes, but sometimes three-quarter-section channel pipes are also used. Use a minimum wall thickness of 225 mm for the inspection chamber, 100 mm thickness for its base and a 75 mm concrete cover for the inspection chamber.
3. Draw the pipe from the inspection chamber to the septic tank using a fall or gradient of 1 in 50. In practice this fall can vary depending on the pipe diameter used. A minimum pipe diameter of 100 mm and a minimum distance of 3.040 m between habitable buildings and the septic tank are recommended.
4. Draw the detail of the septic tank showing the entry and exit points of pipes to inspection chambers at both ends of the septic tank, as shown on page 90. The capacity of the septic tank is calculated based on the number of people living in the building or house. The design formula for septic tanks is $180p + 2000$ litres, where p represents the number of people occupying the building. Note that the minimum capacity of any septic tank is 2.700 m^3.
5. The wall of the septic tank is usually made of 225 mm-thick blockwork or brickwork, on a 100 mm concrete base. The fall of the septic tank is 1 in 20 against the direction of flow.
6. The septic tank should be finished off at the ground level or at 150 mm above ground level (where necessary), with a 75 mm to 100 mm reinforced concrete (RC) slab cover.
7. Continue the pipework from the septic tank to a disposal unit, also called a soakaway pit. Maintain the same gradient all through the length of the pipework. The soakaway pit should be no less than 3.600 m away from habitable buildings occupied by 10 persons, and 9 m if occupied by up to 100 persons.
8. Draw the section of the soakaway pit as shown on page 90, the wall being 225 mm thick. Indicate the hardcore fill up to two-thirds of the depth of the pit. The soakaway pit should be covered with a 75 mm to 100 mm RC slab. The floor of a soakaway pit has no concrete base. However, its walls may be supported on a 100 mm-thick concrete foundation.

TECHNICAL DRAWING 3: BUILDING DRAWING

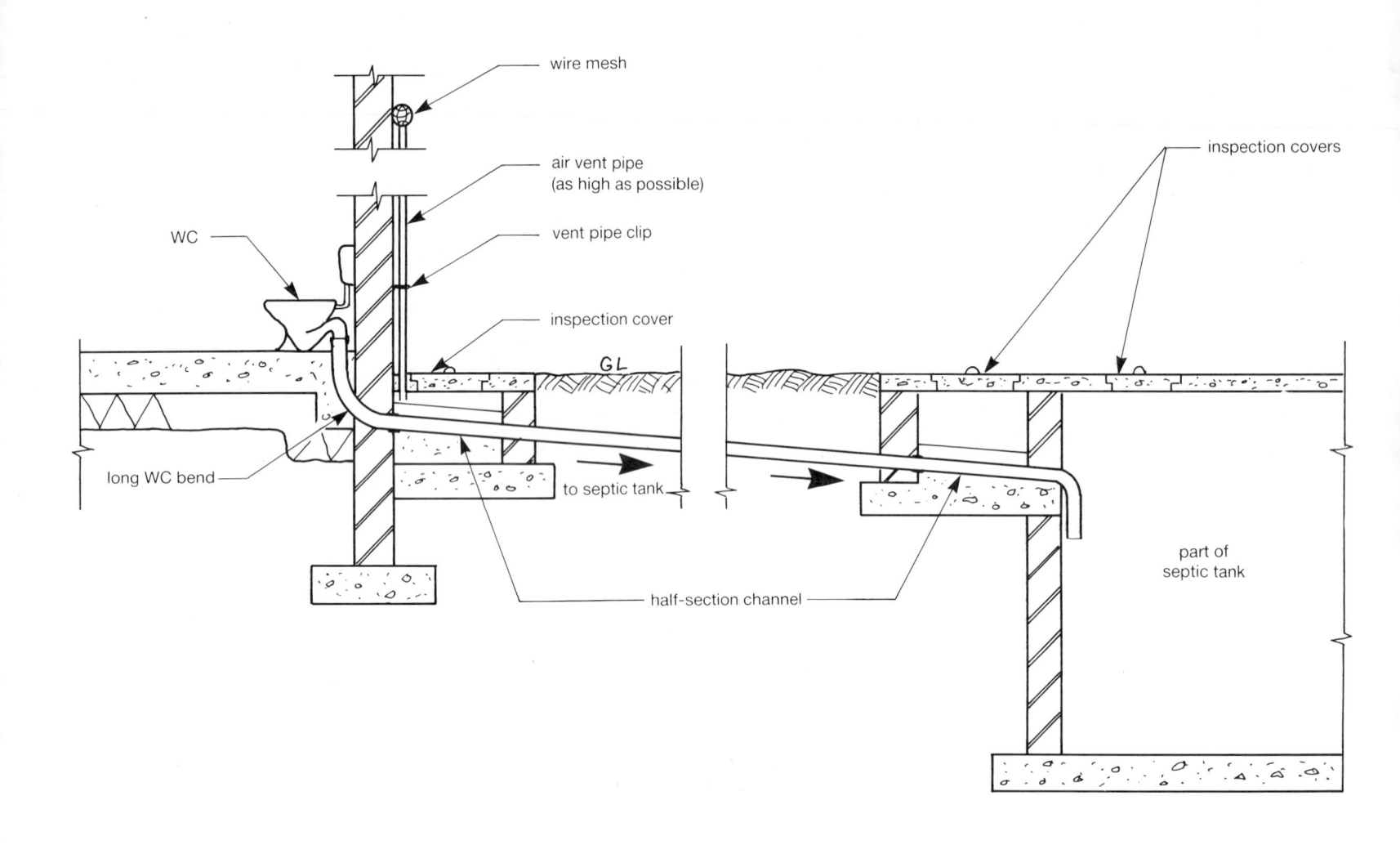

Section through a WC and inspection chamber

88

DRAINAGE AND SANITATION SYSTEMS

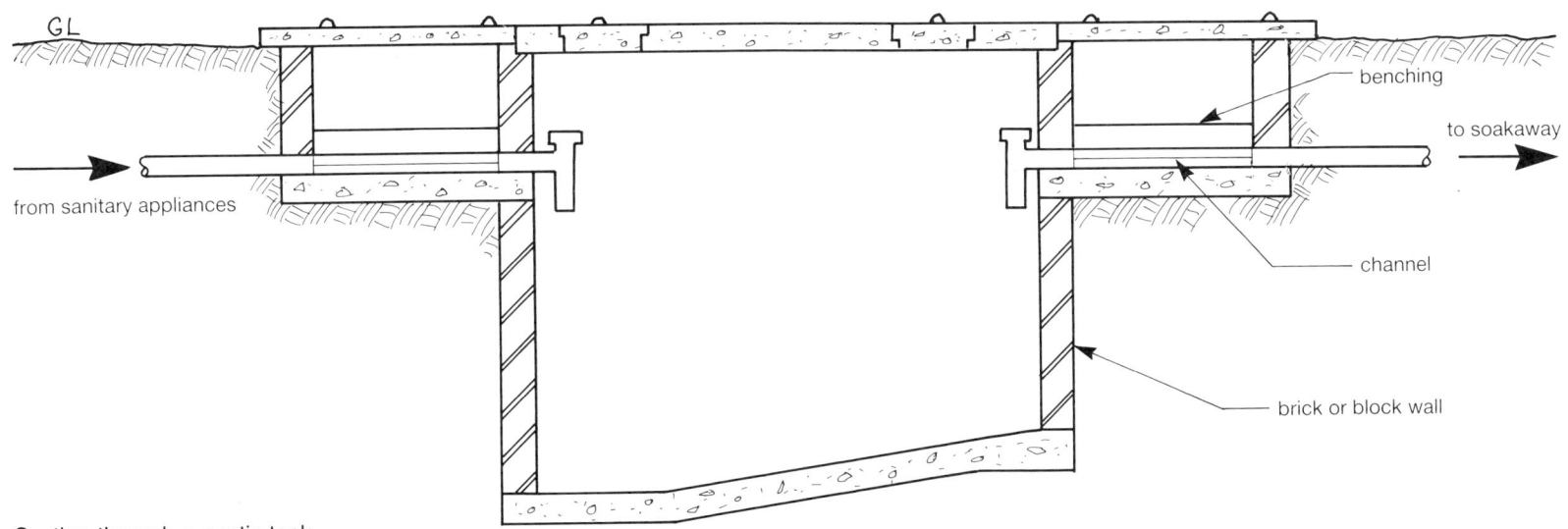

Section through a septic tank

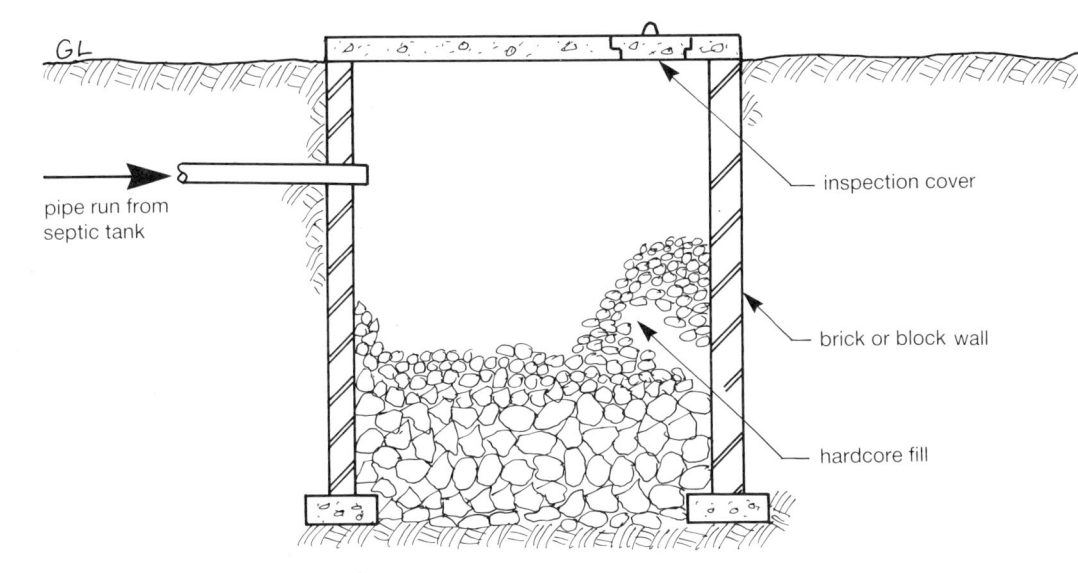

(Labels: GL; from sanitary appliances; benching; to soakaway; channel; brick or block wall)

Section through a soakaway pit

(Labels: GL; pipe run from septic tank; inspection cover; brick or block wall; hardcore fill)

Exercises

1. Design and draw the layout of a cold water supply system showing the following:
   a) the connection of the building main to the town main,
   b) the distribution of water from the building main to the kitchen, bath, and toilet.
2. Draw a layout plan of a hot water supply system.
3. Draw a sectional detail of a septic tank and soakaway pit using appropriate conventions for materials indicated.

89

Index

access road, 24
accuracy of dimensions, 6
adjustable set square, 2
aligned dimensions, 7
alternating current, supply for lamps, 77
angles, 20
    of projection, 20
    dihedral, 20
arches, 43
    axed, 45
    gauged, 44
    rough, 47
arrowheads, 6
asphalt, 71

balance, need for, in sheet layout, 4
balusters, 64
balustrade, 64
bath, 28
bathroom, 28
bedroom, 28
bench mark, 23
borderlines, 5
building construction drawings, 22
bungalow, 28

cabinet oblique projection, 10
carriage, 63
casement windows, 55
central hot water supply, 84
centering, 38
circle, sketching in isometric projection, 14
cistern, 83
clarity of dimensions, 6
clothes closet, 28
code of practice, 81
cold water, designing and drawing a supply system, 81
common rafter, 71
communication pipe, 23, 81
compasses, 2
concrete, 31, 71
    base, 31

determining the size of, 31
floors, 38
roofs, 71
construction lines, 1
contour lines, 23
construction, methods for floors, 34
    monolithic, 34
    separate, 34
conventions, 9
corner title block, 5
corrugated sheets, 71
cross-section, 74
curves, 2
cylinder, sketching of, 15

damp proof course (DPC), 34
damp proof membrane (DPM), 34
datum, 7
deep strip foundation, 32
delayed-action switch, 78
dihedral angles, 20
dimension figure, 6, 7
dimension lines, 6, 7
dimensioning, 6, 7
dimensions, 6, 7
    accuracy of, 6
    clarity of, 6
    completeness of, 6
    readability of, 6
dimmer switch, 78
dining room, 28
direct current, supply for lamps, 77
disposal of waste, 87
distribution board, 79
dividers, 3
dog-leg stair, 62
    drawing of, 68–70
domestic buildings, 81
door, 48
    frames, fixed, 48
    in section, drawing of, 54
    openings, determining the width of, 53
    sizes, 49

doors, 48, 49, 51, 52–3
    classification of, 49
    double, 53
    flush, 51
    match-boarded, 52
    panelled, 49
doors, representing on plans, 52
    in section, 54
    representing in elevation, 53
    sliding, 28
doorways, 25
double-hung sash windows, 55
drainage and sanitation systems, 81
drains, 83
draughtsman, 1
drawing equipment, 1
drawing instruments, 1
drawing pens, 3
drawing sheet, 4, 5

eaves, 71
earthing, 79
electrical safety, 79
electrical details, 22
electrical energy, 77
electrical wiring plans, 77
    components of, 77
    steps in drawing, 78
electrical wiring symbols, 77, 78
electricity in buildings, 77
elevations, 9, 23, 53
    end, 19, 59, 75
    rear, 23
    sectional, 9, 73
erasers, 1
erasing shield, 2
estimates, building construction, 22
examination number, 5
examiners, 5
extension lines, 1, 6

first angle projection, 20
fixed door frames, 48
flat roof, 72

floor, 23
    joists, 35, 37
    plan, 25, 74, 75, 81
    rambling, 74
    slab, 22
floors, 34
    ground, 34
    solid ground, 34
    suspended ground, 34, 35
    upper, 37
fluorescent lamps, 77
flush doors, 49
footpaths, 24, 25
formwork, 38
foundations, 31–3
    deep strip, 32
    depth of, concrete in, 31
    pad, 33
    raft, 33
    strip, 31
    types of, 31, 32
    wide strip, 32
freehand sketching, 10
front elevation, 19, 22, 74
front view, 19
full-length title block, 5

gable end, 71
gables, 75
garage, 28
gauged arches, 43, 44
    to set out on elevation, 44
geometrical stair, 63
going, of a trend, 63
Graphos, 3
ground floors, 34
ground level (GL), 34

H pencils, 1, 10
HB pencils, 1, 10
half-turn stair, 62, 68
    drawing of, 68
hallway, 28
handrails, 67, 70

90

INDEX

hardcore, 35, 87
hardness, of pencils, 1
hatched, lines, 9
hatching, 9
headroom, 64
hip, 71
hipped end, 71
hot water supply systems, 83
   drawing of, 83

in situ oversite concrete, 34
incandescent lamps, 77
Indian ink, 3
industrial buildings, 81
information box, 5
inking, 3
inspection chamber, 87
instantaneous hot water supply, 83
instruments, drawing, 1, 10
isometric grid, 12
isometric projection, 10, 12

jack rafter, 71
jamb, 40

kitchen, 28

L-stair, 61
lamps, 77
landing, 61, 62, 64, 67, 68, 69
layout, sheet, 4
leads, pencil, 1
lettering guides, 3
lettering stencils, 3
lighting fixtures, 77
lighting required in houses, 79
lines, types of, 1
   construction, 1
   extension, 1
lintels, 42, 54, 59
living room, 28
load-bearing walls, 31
local authority bye-laws, 81, 83
louvre blades, 55, 56
louvre windows, 55, 57

main disconnect switch, 77
main switch, 77

margins, 5
match-boarded doors, 49, 52
meter box, 77
metre, as unit of dimensioning, 6
millimetre, as unit of dimensioning, 6
monolithic construction, 34
multistorey building drawing, 25
multiview drawing, 4
muntin, 49

newel, 62, 64
nibs, 3
non-shrinkable subsoils, 31
nosing, 63

oblique projection, 10, 11
office practice, 1
offset section, 26, 27
open-newel stair, 62
open-well stair, 62
orientation, site, 23
orientation symbols, 23
orthographic projection, 9, 19, 20
outlines, 1, 10
overhang, 71
oversite concrete, 34

pad foundation, 33
panelled doors, 49
parapet wall, 73
passage, 28
pencil grades, 1
pencil leads, 1
pencil sharpening, 1
pencils, 1
perspective drawing, 11, 14, 16
   single-point, 11
   single-vanishing-point, 11
   two-point, 11
   two-vanishing-point, 11
physical features, 24
pipes, 81
pitch, 71
pitched roof, 72, 73
pivoted sash windows, 55
plan
   in orthographic projection, 19
   floor, 25

roof, 22, 75
   site, 23
   wiring, 77
plywood, in door construction, 51
plumbing system, 81
   designing, 81
profile planes, 27
projection, 9
   cabinet oblique, 10
   isometric, 10
   orthographic, 9, 19
   planometric, 11, 17
projects, 22
property line, 24
purlins, 71, 76
push-button switch, 78
PVC, 2

quarter-space landing, 61

raft foundation, 33
rafter, 71, 76
rail, 49
raised panel, 49
rambling shape, 76
Rapidograph, 3
readability of dimensions, 6
rear elevation, 22
reinforced concrete (RC) floors, 38
reveal, 40
ridge, 72
rise,
   of roof, 72
   of stair, 63
   calculation of stair, 65
riser, 63
rocker switch, 78
roof, 22
   details in a building drawing, 74
   plan, 22, 75
   classification of, 72
      concrete flat, 73
      flat, 72
      pitched, 72, 73
      timber flat, 72
   components of, 71
room planning and specifications, 28

rough arches, 43, 47
   drawing of, 47
rules, 2, 10
run
   of a tread, 63
   of a roof, 72

safety, electrical, 79
sanitary appliances, 23, 81, 83
sash windows, 55
   double-hung, 55
   pivoted, 55
   sliding, 55
scales rules, 2
screed, 34
sectional views, of buildings, 27
sections, 22
separate construction, 34
septic tank, 25, 87
   and soakaway pit, 87
   drawing of, 87
service entrance, 77
service pipes, 81
set squares, 2
   adjustable, 2
shapes, 14
sheet layout, 4
shield, erasing, 2
shower, 28
sills, 41
single-point perspective, 14
single vanishing point, 14
site, 23
   orientation, 23
   plan, 23
sketching, freehand, 10
   planometric, 15
sketching aids, 12
sleeper walls, 37
sliding sash windows, 55
slope, roof, 72
soakaway pit, 87
socket outlets, 77–79
soils, for foundation, 31, 32
solid ground floors, 34
   steps in drawing, 34–5
spacers, 39
span, roof, 72

INDEX

specifications, 6, 79
squared paper, 12
stairs, 61
  design, 63
  details, drawing of, 64
  in section, procedure for
    drawing, 64–7
  materials, 61
  points to note when designing, 64
  terms used in connection with, 63
  types, 61
    dog-leg, 62
    geometrical, 63
    L-stair, 61
    open-newel, 62
    open-well, 62
    spiral, 63
    straight-flight, 61
staircase, 64
stairwell, 64, 66
steel, 71

stencils, 3
step, in a stair, 63
stiles, 49, 59
storage, 83
string, 63
stringer, 63, 64
strip foundation, 31
survey beacon, 23
suspended ground floor, 35–6
switches, 77, 78
symbols, 9, 77
symmetry, 4

T-square, 2
thatch, 71
third angle projection, 20, 21
three-bedroom flat, 28
three-dimensional presentation, 10
title block, 5
  corner, 5
  full-length, 5

toggle switch, 78
top view, 19
tread, 63
triangular cleat, 75, 76

units in dimensioning, 6
upper floors, 34, 37, 68
utilities, on site plan, 24

valley, roof, 72
vanishing point, 11
vent pipes, 87
verge, roof, 72

wall, 25, 73
  external, 28, 79
  internal, 28, 79
  parapet, 73
wall lighting outlets, 79

wall opening, 40
wall sockets, 77
wall plate, 35
wardrobe, 28
waste, 87
water closet, 28
water supply systems, 81
wide strip foundation, 32
winder, 64
window opening, 58
window sills, 41
windows, 25, 40, 48, 55, 56, 57
  and frame sizes, 55
  casement, 55
  drawing in sections, 59
  locating in walls, 56
  louvre, 55, 57
  sash, 55
wiring plan, 77, 79
wiring symbols, 77
working drawings, 19